D0884495

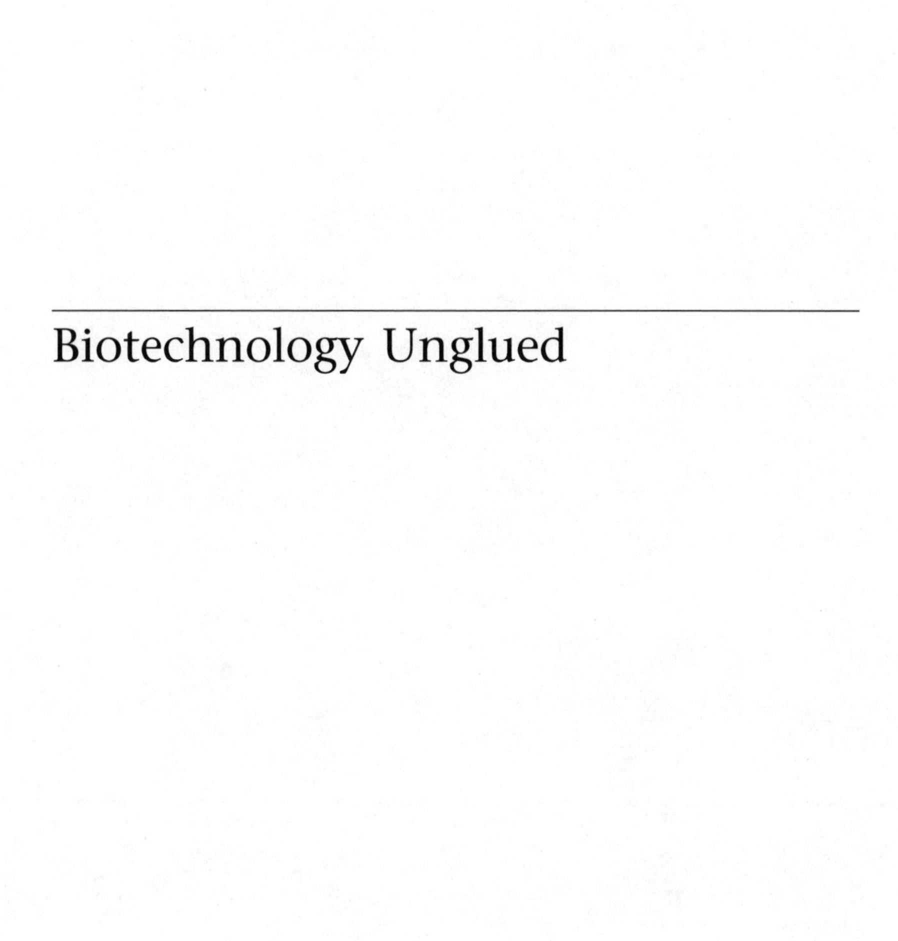

Biotechnology Unglued

Edited by Michael D. Mehta

Biotechnology Unglued: Science, Society, and Social Cohesion

15 14 13 12 11 10 09 08 07 06 05   5 4 3 2 1

Printed in Canada on acid-free paper

Library and Archives Canada Cataloguing in Publication

Biotechnology unglued : science, society and social cohesion / edited by Michael D. Mehta.

Includes bibliographical references and index.
ISBN 0-7748-1133-1

1. Biotechnology – Social aspects. I. Mehta, Michael D., 1965-

TP248.23.B568 2005   303.48'3   C2004-906768-0

Canadä

UBC Press gratefully acknowledges the financial support for our publishing program of the Government of Canada through the Book Publishing Industry Development Program (BPIDP), and of the Canada Council for the Arts, and the British Columbia Arts Council.

UBC Press
The University of British Columbia
2029 West Mall
Vancouver, BC V6T 1Z2
604-822-5959 / Fax: 604-822-6083
www.ubcpress.ca

To Kathy and Kendra –

Birds of a feather flock together

Contents

Acknowledgments

The study of social cohesion is by its very nature an exercise in networking and relationship building. To understand the challenges posed by technologies such as biotechnology to social cohesion requires the input, assistance, and support of many people and organizations. The writing and production of this book comprise a prime example of the necessity of such collaborative efforts.

I would like to thank all of the contributors to this edited book. Your chapters form the foundation of this book. Without your hard work, intellectual devotion, and patience, *Biotechnology Unglued* would not exist.

The financial support of three organizations is especially appreciated. A generous grant from the University of Saskatchewan to support faculty in their publishing endeavours helped to defray a large portion of the costs associated with producing this book. A grant from Genome Prairie's GE3LS program to cover my research on farmers' perceptions of risk and technological change was also drawn upon to cover costs. Finally, a grant from Ag-West Bio Inc. was provided, without strings attached.

I would like to thank Randy Schmidt from UBC Press for his early and continued support of this book. His input and coordination of the many tasks involved in making such a project possible are appreciated. Thanks are also owed to the other staff members at UBC Press for copyediting, indexing, and several of the behind-the-scenes services provided.

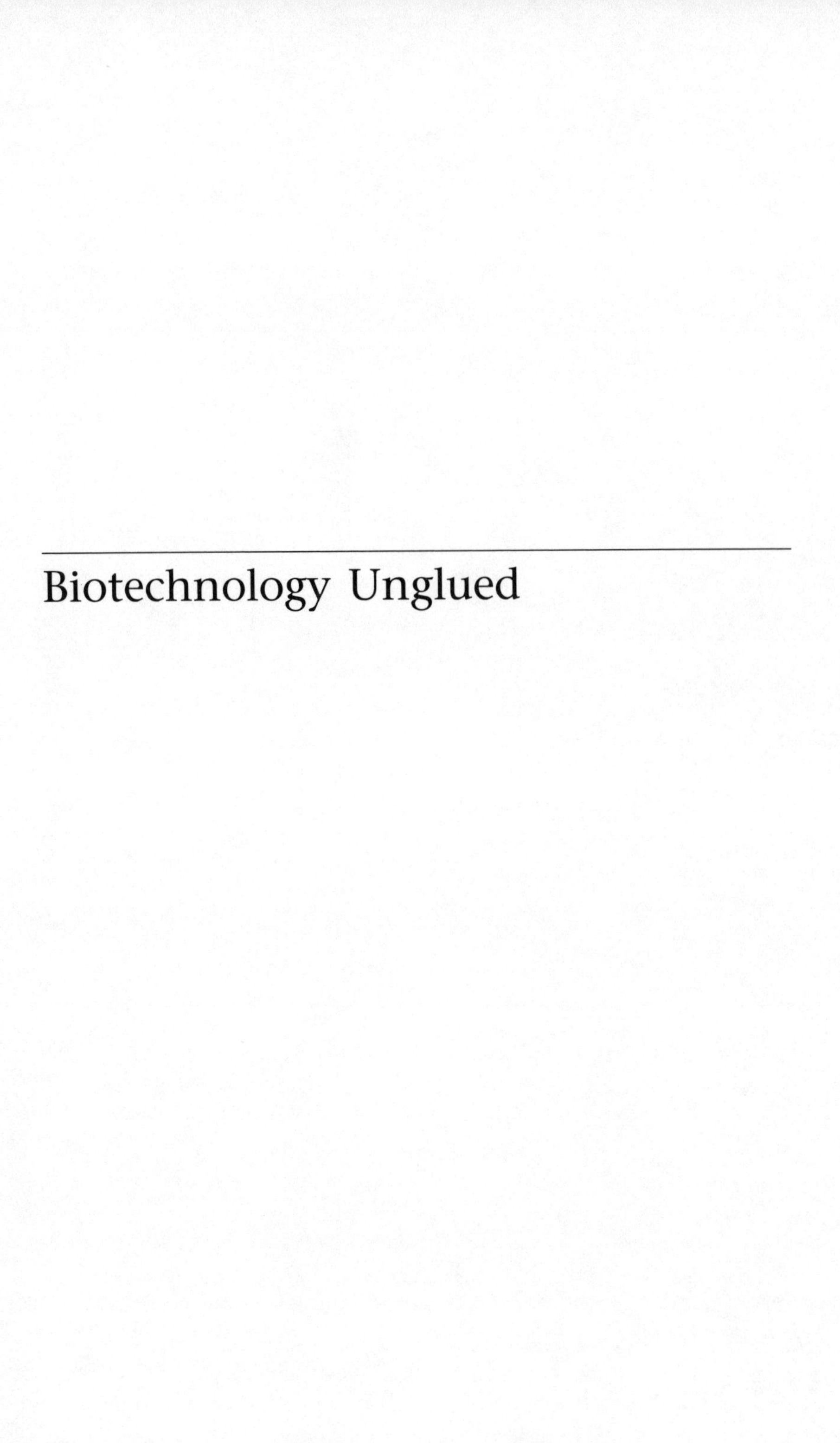

Biotechnology Unglued

1
Introduction: The Impact of Innovations in Biotechnology on Social Cohesion

Michael D. Mehta

Proponents of biotechnology claim that advances in this technology will create a better world: a world free of malnutrition and hunger, with less reliance on herbicides and pesticides, better medical diagnosis and treatment through gene discovery, and more efficient policing and prosecution with forensic techniques using DNA evidence. All of this promise sounds too good to be true. While some innovations in biotechnology provide significant benefits to particular users, the impacts of these technologies on society are often poorly understood. *Can biotechnology threaten the social fabric by weakening, even if temporarily, the social cohesion of society?* This book provides eight case studies on how particular applications in agricultural, medical, and forensic biotechnology affect the social cohesiveness of agricultural communities, citizens of the developed and developing world, consumer groups, scientific communities, and society in general. The glue that holds us together as a society is in danger of dissolving due, in part, to the way that some innovations in biotechnology are currently being developed, marketed, and used.

What Is Social Cohesion?

The literature on social cohesion is rich and varied yet poorly integrated. Social cohesion is a measure of how tightly coupled, robust, and unified a community is across a set of indicators. A community with a strong sense of identity and shared goals is considered to be more cohesive than one without these qualities. A cohesive community is also able to buffer more effectively changes resulting from realignments of international actors, national priorities, local political climates, economic upturns or downturns, and the introduction of new technologies. Recent developments in agricultural, medical, and forensic biotechnology give us a unique opportunity to chart how different communities and actors adjust to the introduction of new technologies and to extend our understanding of the relations between scientific innovations and the social environments into which they are introduced.

There is little agreement on how to define social cohesion. This is somewhat startling considering how widely used this concept is and how quickly some claim that social cohesion has declined in recent years. Moreover, such assertions suggest that social cohesion is a desired state instead of its more likely manifestation as a process that reflects the changing nature of social relations.

Jane Jenson (1998, 3) suggests that social cohesion became popular as a topic of discourse because it illuminates the interconnections between "economic restructuring, social change and political action." Furthermore, Jenson notes that, according to a range of governmental agencies and organizations such as the Organisation for Economic Co-operation and Development (OECD), a cohesive society is assumed to be socially and economically optimal and a decline in cohesion represents a threat to social order. However, it is worth noting that changes in social cohesion are considered to be much more than simply a threat to markets. Judith Maxwell (1996) considers the relationship between social cohesion and the social conditions that indicate when a society fails to function adequately. Maxwell defines social cohesion as the sharing of values that reduce "disparities in wealth and income" while giving people a sense of community (13). It is assumed from this definition that strongly cohesive societies are better able to face the challenges posed by social, economic, and technological change.

This definition demonstrates that social cohesion is a concept that embodies a range of social processes that help to solidify a community and give its members a sense of identity and belonging. Cohesive communities share common values that orient individuals to general needs of the collective but at the same time have the potential to ignore or actively reduce the quality of life for individuals who are in the minority.[1] In other words, by their very nature, cohesive communities are not necessarily more concerned about issues related to things such as equality of opportunity. In fact, some forms of social cohesion may help to reinforce practices that are inherently unjust. As such, the topic of social cohesion plays a significant role in disciplines such as sociology because it provides a lens for observing how social change impacts on community. Cope, Castles, and Kalantzis (1995, 39), following the work of Thomas Hobbes and other social contract theorists, pose the question "in view of the constant competition between human beings for scarce resources, what makes it possible for people to live together peacefully in a civil society?"

Jenson (1998) aids our understanding by providing a glimpse back in time of how the concept of and processes associated with social cohesion evolved theoretically. She points out that social cohesion is a long-standing concern of social scientists and others. In his highly influential book *The Division of Labor in Society* (1893), Émile Durkheim discusses how nineteenth-century Europe experienced decades of social unrest due to rapid changes ushered in by technology, increasing urbanization, and changing gender roles. In his analysis of these changes, Durkheim uncovered the extent to which societies exhibited high degrees of interdependence to satisfy the ever-increasing needs of modernity, where the division of labour required efficient institutions of state and market and presumably some degree of social cohesion.

The extent to which interdependence was linked, both socially and economically, to efficiency is described in Max Weber's book *The Theory of Social and Economic Organization* (1947). Weber's analysis of bureaucracy as a pervasive feature in modern societies demonstrates how a systematic approach to organizing institutions such as government along formal channels of responsibility meets some of the needs of an interdependent and diverse society. When the role of bureaucrat is socially sanctioned , support for an elaborate hierarchical division of labour is assured. For Weber, the ideal bureaucrat was someone who could apply impersonally the rules of an organization in a professional manner. The bureaucrat owns neither the means of production nor the means of administration and is viewed as an agent or supporter of rational administration. In this sense, the bureaucrat maintains the efficient functioning of an administrative machine by oiling the mechanisms that support the division of labour made necessary by modernity. In a society where class struggle and the existence of inequalities are systemic, the functioning of bureaucracies, and the changes occurring institutionally as a result of capitalism's unfolding, bring the topic of social cohesion to the foreground.

Weber viewed bureaucracies as highly efficient ways of formally entrenching rational analysis in the goals that people and societies hold. Consequently, Weber viewed bureaucracy as inescapable. While he saw the process of bureaucratization as inevitable, at best he was ambivalent about this development. Nonetheless, many modern-day states formally and informally create bureaucratic procedures that are designed to weed out inefficient practices and to stimulate the development of innovations that improve profitability under the assumption that economically successful societies have happier, more cohesive, and more productive citizens. To wit, a society where the production of wealth is encouraged, and where some means for rationally allocating wealth exists, is more likely to place a premium on strong social cohesion, especially if such cohesion welds workers and management together in the name of efficiency, profitability, and global competitiveness.

Talcott Parsons did not share this optimism and faith in the value of socially diverse institutions and their positive impacts on social cohesion. In *The Structure of Social Action* (1937), Parsons discusses the limits faced by capitalist societies for creating the social conditions necessary for ensuring social stability. For Parsons, the nation-state is the best form of social organization for ensuring that social, political, and economic needs are met under capitalism. Clearly, Parsons was aware of the threats posed by excessive liberalism and he treated society as a system that was unified by shared values. A society without stability is one where individual behaviour, especially in the marketplace, fails to provide the corrective forces necessary for social order. Parsonian functionalism emphasized the need to depathologize poorly functioning institutions, and markets, by striving for consensus rather than diversity of opinion. With many changes occurring in society during the 1930s and 1940s, a Parsonian approach to managing society, by controlling its systems and subsystems, garnered criticism. Rapid social change and an increased awareness of the collateral damage associated with rapid industrialization began to demonstrate that shared loyalties and values, within social systems characterized by interdependence, were not necessarily socially optimal.

This systems approach also emphasized the primacy of institutions over individuals. This emphasis became problematic when it was realized that even the most stable and historically cohesive societies continued to function without manufactured consensus. Jenson (1998, 10) observed that supporters of the Parsonian tradition reluctantly "came to realize that consensus was not necessarily a requisite of cohesion and that conflict could be healthy."

We can more completely understand why conflict emerges in societies that are, on most levels, considered cohesive by returning to Durkheim for assistance. Durkheim (1893) used the term "anomie" to describe a condition in society where people break the rules that are normally used to create predictable behaviour. Anomie is a breakdown of social norms and the conditions required for their enforcement and stability. For Durkheim, sudden changes in society generated anomie by challenging how people think and behave and by weakening social bonds. These changes are expected to weaken group-oriented goals and to coincide with periods of economic dislocation and higher rates of crime and deviance.

Durkheim proposed that societies evolved along a path moving from simple to more complex forms. He labelled the simple stage *mechanical* and the more complex stage *organic*. In the mechanical stage, a society contains people who share common roles and perform group-oriented tasks. Mechanical societies are relatively less formally organized and are flatter hierarchically. In contrast, an organic society is more complex and exhibits specialization in terms of social roles and division of labour. Organic societies are characterized as being more impersonal and less stable. Conditions that generate anomie are expected to have more impact on organic societies. Such societies are also characterized as having more need for rational administration through formal bureaucracies so that the complex relationships that exist in such societies can continue to function smoothly and efficiently.

Many developed countries may be defined as organic in both nature and functioning, and as such they place a strong emphasis on social cohesion as a cultural resource (Berger 1998). This status creates a range of stresses that sometimes pit social cohesion against the very nature of pluralism itself. A vision of what constitutes a cohesive society should not include a desire to return to supposedly homogeneous communities of the past. Organic societies must recognize differences and the benefits that can be derived from diversity. However, such a condition requires that public and private institutions respond to the challenges posed by dynamic actors and agents. It is important to note that social cohesion does not imply social, political, or even economic stability. In some hidden way, change seems to foster different kinds, or degrees, of cohesion. A "healthy community" is one that can grow from within by responding to change without becoming overly introspective or, worse yet, overly corrective. Here the role of institutions, and their legitimacy, becomes key.

A cohesive society is one that can sense itself and, more importantly, mobilize the necessary "social capital" to correct distortions.[2] It contains institutions that foster trust and commitment that work in the interest of the public good. Such a society utilizes social capital by actively building networks and forums that encourage public debate. Indeed, such societies welcome the opportunity to explore complex and perhaps even divisive issues such as those raised by the introduction of biotechnology. By taking advantage of social capital, a society becomes more responsive to the inequities generated by neoliberal economic policies that sometimes work in the direction of weakening or shifting the nature of social cohesion.[3] From a Durkheimian perspective, building social capital helps to reduce the degree of anomie that may follow rapid changes in society.

Some of the challenges that confront our society seem to be overwhelmingly large. Diane Bellemare and Lise Poulin-Simon (1994, 12) note that particular economic and social policy decisions made recently in Canada have potentially negative impacts, including increased rates of unemployment and growing disparities in income that "contribute to social division and the erosion of social cohesion." If we combine this observation with that made by Thomas Homer-Dixon (1994), we may see why some are quite concerned about the future. Homer-Dixon is well known for his work on the environmental basis of conflict. For him, scarcity of things such as food and drinking water can precipitate incidents of violence in poorer countries that "directly challenge the national security interests of developed countries, including Canada" (7). In this instance, social cohesion in developed countries can manifest itself in the form of trade agreements or military pacts that help to minimize such risks. Additionally, international agreements such as the Kyoto Agreement on the reduction of greenhouse gas emissions and the Convention on Biological Diversity may become part of this global chess game. However, from an equity and humanitarian basis, such practices are questionable.

Paul Bernard (1999) addresses this issue indirectly by providing three broad sets of responses to the question posed by Cope, Castles, and Kalantzis (1995) on how people can continue to live together peacefully under conditions of scarcity and competition. First, Bernard suggests that a classical liberal approach to this question defers to the wisdom of the marketplace by assuming that individual preferences are expected to "produce a flexible and viable social order from the apparent disorder of individual freedoms" (4). Second, some may believe that placing a high degree of faith in the "invisible hand" of the marketplace fails to prevent what Bernard refers to as the "unchecked pursuit of individual advantages" (4). This concern is echoed by Gordon Betcherman and Graham Lowe (1997, 42), who state that "the current of economic life is now running against the collectivity of communities. We are at risk of becoming a society of consumers and customers, not citizens." Betcherman and Lowe observe that the creation of greater social inequities heightens the "individualization of risk," where individuals are expressing an interest in exerting their rights to disconnect from services provided to the collective in favour of self-directed services (e.g., two-tiered medicine).[4] To minimize this risk, it is necessary to strengthen institutions that reflect and support widely shared values and to give such values a moral underpinning. Third, others believe that it is necessary to root out the social bases of injustice by minimizing the inequities that exist in the social order by class, sex, ethnicity, race, sexual orientation, and so on. The presence of widely experienced forms of injustice may lead to social conflict and correspondingly affect social cohesion.

From this review, it should be evident that much of the theorizing on social cohesion has formally, or tacitly, an emphasis on the importance of ensuring that equity concerns are addressed. It is essential to reiterate that social cohesion can work to ensure that injustices are minimized, or, conversely, can support the conditions that allow injustices to remain entrenched or even to develop more markedly.[5] Many of the debates over innovations in biotechnology pick up on this thread.

Social Cohesion and Biotechnology

In the summer of 2001, I sent out a call for papers on the social impacts of biotechnology to several individuals, institutions, and e-mail-based listservs. The chapters in this book represent a wide range of perspectives on how advances in agricultural, medical, and forensic biotechnology may threaten the social cohesiveness of different kinds of communities and at different scales. Each chapter represents a case study on how the development, regulation, commercialization, or use of different innovations in biotechnology affects the social cohesiveness of agricultural communities in the developed and developing world, consumer groups, scientific communities, and society in general.

In Chapter 2, I examine how the introduction of innovations in agricultural biotechnology (also known as ag-biotech) influences the social cohesion of food producers (farmers). Specifically, this chapter examines how the use of herbicide-tolerant (HT) canola and genetically modified corn (Bt) affects land tenure, management practices, and the social fabric of agriculture in western Canada. I address a number of questions. Does the use of HT crops provide an advantage to large-scale farmers relative to small-scale farmers? Does the use of this technology affect the cohesiveness of agricultural communities by creating a culture of surveillance (e.g., to ensure that Technology Use Agreements are followed)? Does the use of this technology deskill farmers? Can innovations in agricultural biotechnology stimulate conflicts between farmers? This chapter demonstrates how innovations in agricultural biotechnology may threaten social cohesion, and it suggests that weakly cohesive communities are more likely to suffer when economic fortunes decline and are much less capable of mobilizing the social capital needed to sustain themselves and to be innovative. As a result, weakly cohesive agricultural communities represent a decline in the quality of living in rural communities.

In Chapter 3, Jacqueline Broerse and Joske Bunders deconstruct the argument that biotechnology is needed to stave off global famine. Biotechnology is often presented as a potentially powerful factor in contributing to poverty alleviation, food security, and sustainable development in developing countries. When we look at the innovations currently being developed through biotechnology research and development (R&D), we can, however, conclude that these innovations are usually inappropriate for this purpose. Skepticism therefore prevails in the development community about the usefulness of biotechnology as an instrument for achieving these goals. In this chapter, Broerse and Bunders provide a broad overview of a range of recent developments in agricultural biotechnology primarily focusing on the developing world. They argue that, if biotechnology is to benefit the poor, a double shift in the research paradigm is needed. Biotechnology R&D should be specifically (1) focused on agro-ecological systems and products important to poor people and (2) contextualized within the broader socioeconomic and cultural situation of the poor while fostering a deeper understanding of sustainability issues. They conclude that implementing an interactive and participatory approach to the biotechnological innovation process – involving farmers, scientists, and other stakeholders as well as enhancing a broader process of training of human resources and institutional change – is the way to proceed in the field of biotechnology development for small-scale, resource-poor farmers. Unless such an approach is taken, the introduction of new ag-biotech innovations is likely to threaten the stability and social cohesiveness of the developing world by jeopardizing food security and reducing the viability of small-scale farming.

In Chapter 4, Christopher Vanderpool, Toby Ten Eyck, and Craig Harris examine how the introduction of genetically modified foods into the US marketplace has created a crisis of legitimation that serves to weaken trust in regulatory agencies and the food industry. Although genetically modified foods are novel, at first glance they are not more novel than other agrifood technologies. Yet genetically modified organisms (GMOs) have become embroiled in highly rancorous conflict in many societies. Vanderpool and colleagues suggest that this situation has come about because (1) genetic engineering (GE) alters the nature of food in essential ways; (2) especially in the United States, GE foods were introduced in a context of growing concerns about food safety; and (3) agrifood factions interested in GMOs attempted to keep public awareness of GM foods and their various attributes as low as possible. These three factors have not only produced divisive conflicts over agrifood biotechnology itself but also caused a legitimation crisis wherein significant segments of society begin to withhold their allegiance to the state and become less accepting of government claims and regulations.

In Chapter 5, Margareta Wandel documents how consumers in Norway have responded to the introduction of genetically modified foods and how their perceptions of the risks posed by these foods have been addressed. Consumer studies show that Norwegians are concerned about the use of gene technology and that negative attitudes are particularly strong when this technology is used in food production. This chapter takes as a point of departure the statements and conclusions made in two lay panel conferences, carried out in Oslo, and builds on Ulrich Beck's (1992) work on the "risk society." In short, Wandel's analysis demonstrates the need to understand how consumer perceptions of food are linked to the trust that individuals have in institutions. In situations of low trust, concerns about food safety are likely to be amplified. One remedy explored by Wandel for building trust involves the mandatory labelling of food containing ingredients from genetically modified sources. Since trust is an essential component of a socially cohesive society, important lessons from the Norwegian consultation exercise become apparent.

In Chapter 6, Kyle Eischen explores the social questions raised by the Iceland Genetic Database. On one level, there are serious issues of privacy, competition, commercialization, and individual rights that challenge or extend existing local legal codes and social norms in fundamental ways. On a higher level, the developments in Iceland provide a way to outline how global economic, social, and technological trends shape and connect with local resources, needs, and policies. The development of the Icelandic database depends on a unique set of circumstances that have given Iceland an extraordinary degree of cultural, biological, and social cohesion. In combination, they provide for a unique data set that is invaluable to new global industries such as medical biotechnology. Eischen suggests that information technologies, as both process and product, build on and embody this existing social knowledge and thus represent the construction of new social norms and institutions in unforeseen ways. As such, the construction of new local social structures is intimately tied to broader global trends. Simply, Iceland matters not only because of concerns over privacy or commercialization of individual genetic information on a regional level but also because the debate itself exists only when broader global trends impact in real ways on specific regions and populations. Iceland offers a detailed example of how social cohesion is realigned with powerful and sudden changes in technology outside the immediate control of individuals or existing institutions. How societies such as Iceland respond to these transformations, how they use existing social capital and resources within these new global relationships, is a central feature structuring future social cohesion.

In Chapter 7, Neil Gerlach focuses our attention on how biotechnology intervenes in bodily processes by facilitating new modes of surveillance. Canadians seem to be relatively unaware of DNA matching and banking within the criminal justice system. Since 1995, Canadian police have been empowered to obtain warrants to seize DNA samples from suspects of crime, and since 2000 a fully operational national DNA data bank has been in place to store DNA samples of convicted offenders. A number of interest groups have raised concerns about the potential of these technologies for intruding upon privacy rights of citizens as set out in the Canadian Charter of Rights and Freedoms. Nevertheless, DNA testing and banking remain uncontroversial in the public sphere. Gerlach describes the entry of DNA testing and the DNA bank into Canadian criminal justice and examines the enabling conditions that have allowed these modes of surveillance to enter Canadian society in a relatively unproblematic way. Specifically, fear of crime, increasing comfort with surveillance, redefinition of criminality as based in nature rather than nurture, and rationalization of criminal justice institutions using corporate models of efficiency have allowed for an easy normalization of DNA testing and banking. The result has been a growth in state powers of social control and less emphasis on human rights in the execution of criminal justice. In short, technologies of surveillance are more likely to be tolerated, or even embraced, when beliefs about the collapse of formerly socially cohesive societies are widely held.

In Chapter 8, drawing upon literature usually reserved for collections of art and natural science during the rise of the industrial North-West, Annette Burfoot and Jennifer Poudrier explore several aspects of biotechnology as collection. The chapter begins by reviewing the history and critical views of collection. Collecting principles (social progress, appropriation, classification and display, and value) are applied to contemporary biotechnology collections in the second section of this chapter. The modern collections include those involved in plant bioengineering, plant and animal preservation, the Human Genome Project, pharmaceuticals, and medical biotechnology. Burfoot and Poudrier claim that the main principles of modern collection are the decontextualization and appropriation of genetic variation and the designation of value through patent protection. As such, corporate control of plant genetic resources and other kinds of DNA has profound effects on indigenous cultures and peoples of the developing world. Corporate practices such as the race to sequence and patent parts of the rice genome intensify a monocultural ideology in which local self-sufficiency and environmentally sound practices are eclipsed by the forces of a globalized food marketplace. In short, the practice of collecting DNA for corporate gain accelerates social change in many parts of the world by introducing new values such as efficiency, scientism, and Western conceptions of intellectual property that may affect the social cohesiveness of communities in the developing world.

In Chapter 9, Robert Dalpé, Louise Bouchard, and Daniel Ducharme examine the new research dynamic in biotechnology generated by industry's direct relations with university or public laboratory researchers. Their objective is to determine how researchers act and respond to their new environment and to understand the nature of constraints imposed. Dalpé and colleagues present results from a case study dealing with the discovery of two genes associated with breast and ovarian cancer, BRCA1 and BRCA2. Their analysis is based on in-depth interviews of seven researchers exemplifying the different profiles in research dynamics. They conclude that firms and public organizations engage in frequent conflicts concerning patenting of both genes and their subsequent industrial applications. Their data suggest that scientific collaboration is more difficult when intellectual property is an issue and that such conflict may weaken the social cohesiveness of scientific communities and create a sense of distrust.

The chapters in *Biotechnology Unglued* show how a range of applications derived from the science of biotechnology affects in manifold ways the social cohesiveness of different kinds of societies. These impacts may be regional and sectoral in nature, as in my chapter on the introduction of genetically modified canola and corn into agricultural communities in western Canada; national in scope, as in the chapters by Vanderpool and colleagues, Wandel, Eischen, and Gerlach; global in nature, as demonstrated by Broerse and Bunders and Burfoot and Poudrier; or specific to particular scientific communities, as illustrated by Dalpé and colleagues. Finally, each chapter in this book strives to show the two faces of biotechnology by exposing the promises and perils associated with a range of innovations, and it demonstrates how particular kinds of technology-society and technology-corporate configurations affect social cohesion by creating cultures of surveillance, competition, social exclusion, and control. While advances in biotechnology continue to be made in laboratories around the world, a significant social experiment is occurring simultaneously. Will these new technologies unglue, or perhaps realign, the social fabric as we know it? Clearly, this book is only a starting point for investigating the impacts of technology on social cohesion.

Notes

1 The attacks on New York City and Washington, DC, on 11 September 2001 illustrate this point. The "War on Terrorism" has reinvigorated patriotism in the United States. The sale of American flags and the general support for military action against the Taliban government of Afghanistan represent a new kind of social cohesion aimed at weeding out an identifiable threat (terrorists in general, Osama Bin Laden in particular). An unfortunate side effect of this new cohesion is the discriminatory actions taken against many visible minority groups through measures such as racial profiling at airports and border checkpoints.

2 According to Robert Putnam (1993, 2000), social capital is defined as networks, norms, and trust that operate within social organizations to facilitate mutual benefit. Organizations that can leverage social capital are considered to be cooperative and collaborative in nature and structure and to formally and informally support exchanges of information and expertise.

3 Neoliberalism is a philosophy that assumes that the marketplace is the most efficient way of dealing with social problems. It values maximizing transactions on a global scale, formalizing exchanges, and creating an "audit society." Critics of neoliberalism claim that such policies increase the gap between rich and poor and benefit powerful financial institutions that enjoy support from organizations such as the International Monetary Fund (IMF) and the World Bank. Recent protests at various meetings of the World Trade Organization (e.g., Seattle and Quebec City) illustrate how social cohesion works on a different level. Protesters, and many of their supporters who did not physically take part in the actions, represent a relatively cohesive community who share common concerns and values about the changing nature of international trade. On some levels, the concerns of these protesters represent a direct critique of neoliberalism.

4 To a certain extent, this trend is already well established. Private insurance coverage and self-directed pension plans are common examples of how individuals build secondary sets of safety nets.

5 Even Second World War Nazi Germany was a relatively cohesive society. Cohesion is by no means a guarantee that systemic discrimination and state-sponsored genocide cannot happen. I attribute this observation to Professor Peter Li from the Department of Sociology, University of Saskatchewan.

References

Beck, U. 1992. *Risk Society: Towards a New Modernity*. London: Sage Publications.

Bellemare, D., and L. Poulin-Simon. 1994. *What Is the Real Cost of Unemployment in Canada?* Ottawa: Canadian Centre for Policy Alternatives.

Berger, P. 1998. *The Limits of Social Cohesion: Conflict and Mediation in Pluralist Societies: A Report of the Bertelsmann Foundation to the Club of Rome*. Boulder, CO: Westview.

Bernard, P. 1999. *Social Cohesion: A Critique*. Ottawa: Canadian Policy Research Networks.

Betcherman, G., and G. Lowe. 1997. *The Future of Work in Canada: A Synthesis Report*. Ottawa: Canadian Policy Research Networks.

Cope, B., S. Castles, and M. Kalantzis. 1995. *Immigration: Ethnic Conflicts and Social Cohesion*. Sydney: NLLIA Centre for Workplace Communication and Culture.
Durkheim, É. 1893. *The Division of Labor in Society*. New York: Simon and Schuster.
Homer-Dixon, T. 1994. "Environmental and Demographic Threats to Canadian Security." *Canadian Foreign Policy* 2. On-line at <http://www.carleton.ca/npsia/cfpj/>.
Jenson, J. 1998. *Mapping Social Cohesion: The State of Canadian Research*. Ottawa: Canadian Policy Research Networks.
Maxwell, J. 1996. *Social Dimensions of Economic Growth*. Eric John Hanson Memorial Lecture Series, Volume 8. Edmonton: University of Alberta.
Parsons, T. 1937. *The Structure of Social Action*. New York: Macmillan.
Putnam, R. 1993. "The Prosperous Community: Social Capital and Public Life." *American Prospect* 4 (13). On-line at <http://www.prospect.org/print/v4/13/Putnam_R.html>.
–. 2000. *Bowling Alone: The Collapse and Revival of American Community*. New York: Simon and Schuster.
Weber, M. 1947. *The Theory of Social and Economic Organization*. New York: Simon and Schuster.

2

The Impact of Agricultural Biotechnology on Social Cohesion

Michael D. Mehta

Many conventional farmers in the developed world have been quick to adopt agricultural innovations stemming from biotechnology.[1] Since the mid-1990s, farmers in the United States, Canada, and Argentina have switched over to growing large quantities of genetically modified corn, soybean, and canola.[2] These three countries produce a significant amount of the world's supply of food from genetically modified plants. The move from growing crops based on conventional farming techniques to using genetically modified plants has generated new sets of interpersonal and institutional relations that affect social cohesion.

The literature on social cohesion gives us several ways to examine how the introduction of new technologies influences the cohesiveness of farming communities. Farming is anecdotally considered the "oldest profession" and is often linked historically to nomadic communities that became more geographically fixed and agriculturally based. Our understanding of social cohesion has been influenced by this transition. Geographically fixed communities that are tied directly to the means of production, the land, are expected to have an anchor that generates cohesion by facilitating shared experiences. In contrast, nomadic communities are expected to have different points of contact along which cohesion is based. According to Émile Durkheim's (1893) concept of anomie, agriculturally based communities structure their activities along both natural cycles (e.g., climate, pests, soil quality) and farm management practices (e.g., crop rotation, harvesting). Sudden, or anticipated, changes in climate or farm management practices may create a state of anomie. Fluctuating market prices or changes in consumer or processor demand for agricultural products can also affect the functioning of farms and the cohesiveness of farming communities. Even increases in the price of fuel create strains with which farming communities have to deal. Rising fuel prices increase the cost of fertilizers (which are often petroleum based), the transportation of goods, and the day-to-day functioning of farms.

Nomadic communities probably use mobility to reduce anomie by relocating themselves both physically and socially. Physically, such communities can take advantage of better access to resources such as food and water, while social mobility may offer options for reduced inter- and intra-community conflict. It is important to note that neither a nomadic nor an agricultural community is necessarily superior in terms of cohesiveness. However, there are unique dynamics associated with both kinds of communities that are worthy of further consideration. Due to the scope of this chapter, we now turn our attention to agricultural communities like those found in many parts of western Canada.

In the Canadian provinces of Manitoba, Saskatchewan, and Alberta, canola is the flagship crop of choice for genetic modification. Although genetically modified flax[3] and corn are grown on the Prairies, canola represents a significant investment in this technology.[4] Derived from rape seed, canola was created by Canadian scientists working at the University of Saskatchewan and Agriculture Canada in the 1950s. Since that time, several of the properties of canola have been modified both to improve the safety of this product and to enhance properties associated with a healthy diet.[5] Because of these characteristics, canola oil is used extensively in a wide range of processed foods and is a major export crop for Canada. However, the story of canola would not be complete without looking at the other side of the coin: namely, how has the transition to genetically modified canola affected the cohesion of farming communities?

To answer this question, we need to assume that particular characteristics of a product represent unique sets of relations between farmers and others in the food chain. Additionally, some of these characteristics may affect relations between farmers themselves. In other words, how does genetically modified canola differ in this respect from nongenetically modified canola?

There are two popular varieties of genetically modified canola grown in the Prairies: Monsanto's Roundup Ready™ and Aventis's Liberty Link™.[6] Both kinds of canola are genetically modified to tolerate exposure to broad spectrum herbicides sold by each company. The Roundup Ready variety is tolerant of Monsanto's highly profitable herbicide Roundup.[7] Glyphosate is the chemically active ingredient in Roundup. It kills most vegetation that it comes in contact with while sparing those plants that are genetically modified to tolerate it.

The Liberty Link system works in a similar fashion but relies on the insertion of genes into canola plants that confer resistance to the herbicide glufosinate ammonium. Both herbicides are used by Prairie farmers to "burn off" fields prior to planting and during the growing season to control weeds. Aventis claims that "Liberty Link is a production system where seed and herbicide work together to deliver clean fields and higher yields."[8] For farmers, the use of these technologies represents lower input costs and less labour.

Instead of spraying a field with a broad spectrum of herbicides throughout the growing season, farmers only need to spray once or perhaps twice. On the other hand, agreements between biotechnology producers and seed companies usually fix the price of the selective herbicide and its sourcing.[9]

Many concerns about the use of herbicide-tolerant crops have emerged in recent years. Most of these concerns focus on the potential impacts that such crops may have on human health and the environment.[10] In contrast, little discussion of the social impacts of these applications is heard. Does the use of herbicide-tolerant crops provide an advantage to large-scale farmers relative to small-scale farmers? Does the use of this technology affect the cohesiveness of farming communities by creating a culture of surveillance to ensure that Technology Use Agreements are followed?[11] Does the use of this technology deskill farmers? Can innovations in agricultural biotechnology stimulate conflict between farmers?

Do Large-Scale Farmers Benefit More from Innovations in Agricultural Biotechnology?

Herbicide-tolerant crops represent a valuable tool for farmers interested in better control of weeds. Early adopters of this technology were able to take advantage of a relatively weak regulatory environment in Canada and very low levels of public awareness.[12] Since most applications involving the insertion of genes to confer herbicide tolerance in plants did not reach the Canadian food consumer until 1996, farmers who planted these crops in 1995 were rewarded more handsomely for taking this risk. Ostensibly, larger-scale farmers are better positioned to take such risks. They are more likely to grow a range of crops, have access to cutting-edge technology, can afford more help, and have a land base to draw on as equity to buffer themselves from variations in the market for the sale of their produce. Additionally, since many food processors source raw materials directly from farms under contract, it is likely that having a contract with a processor protects larger-scale farmers from market price fluctuations. The formal and informal relationships that many large-scale farmers have with the food-processing industry probably instill in these farmers a sense of confidence when it comes to trying new technologies such as herbicide-tolerant crops.

To understand how new innovations in agriculture impact on social relations of production, consider the work of rural sociologists Rogers (1962); Buttel, Larson, and Gillespie (1990); and Fliegel (1993). Early work in this area concentrated on the adoption of new agricultural technologies by farmers. Emphasis was on the "cultural lag hypothesis" and the discovery that diffusion was a patterned process. At this time, diffusion research was characterized by a linear, sociopsychological approach. Researchers attempted to understand farmers' adoption behaviour and to determine characteristics responsible for individual propensity to be innovative. Key to this understanding was the assumption that agricultural technology was economically beneficial to the farmer and that he or she was free to adopt it or not. The farmer was seen as a rational actor who, under some circumstances, was influenced by social and cultural trends. The transmission of information about agricultural innovations and the role of the media in shaping adoption behaviour were early interests of researchers in this area.

This model of diffusion represents a linear, top-down approach to the adoption of new agricultural technologies. New technologies are developed, farmers eventually adopt them, and those who take the risks associated with innovation are rewarded. Such a pattern existed in conventional agricultural systems. New developments in agricultural biotechnology have changed the nature of this pattern of diffusion. Now more than ever, sociopolitical factors influence the ways in which farmers respond to new agricultural technologies. Individual farmers may resist adopting a new technology for a variety of reasons, including concerns about the multinationalization of food production, the role of agricultural subsidies, the privatization of research, international trade agreements, and weak acceptance by consumers.

Farming communities are changing because of the restructuring of the food industry and these new relationships. Farmers who grow genetically modified crops, or who enter into agreements with seed companies or food processors, may find themselves in an oligopolistic market where there is only one buyer for what they grow. Sometimes this buyer belongs to the same conglomerate that provides seed and supplies. This dynamic has led to a concern that farmers are becoming "bioserfs" or "DNA pawns." On a different level, these changing relations are likely to affect the cohesiveness of farming communities. In the same way that the practices of organic farmers sometimes clash with those of conventional farmers, the introduction of a third wave of farming involving the use of genetically modified crops may create even greater challenges to community building and solidarity.[13] These approaches represent more than different philosophies of farming; they symbolize a range of social and cultural values too.[14]

On a global level, small farms are encountering increasingly desperate conditions. Due to low commodity prices, reduction of subsidies domestically but not internationally, increases in the prices of fossil fuels, and restructuring of the global food industry, small farms are disappearing. On the surface, large monoculture operations appear to be more productive, thus justifying the death of small farms. However, small farms employing traditional polyculture use the land more intensively, utilize resources more efficiently, and can produce more food overall per acre.[15] This observation suggests that small farms are multifunctional and can better serve the needs of rural communities. Small farms are more likely to embody a diversity of ownership, landscapes, cropping systems, cultures, and traditions. They are more likely to stimulate what Aldo Leopold (1949) characterized as a "land ethic," where stewardship of natural resources and investments in conserving biodiversity are more likely.[16] Small farms may facilitate the empowerment of local community by stimulating a greater sense of personal responsibility for the land and the communities that are connected to it. Along with greater diversity, communities that have a more equitable distribution of land have access to more social capital. Finally, small farms are more likely to improve the personal connection that consumers have with farming in general and food in particular. Communities with small-scale farming hold farmers' markets and sometimes engage in direct marketing strategies. In short, innovations in agricultural biotechnology are geared toward a transnational flow of goods in which food processors can source ingredients for processed foods from any source or country they wish. This restructuring of the food industry is profoundly at odds with the values and benefits associated with small-scale farming.

Is a Culture of Surveillance Being Created in Farming Communities?

One of the features of rural life that appeals to many is low-density living. Open spaces, with large expanses of land, are the most visible difference between rural and urban landscapes. A certain degree of privacy and autonomy is often expected to follow the contours of the land. In recent years, new threats to privacy have emerged in farming communities. The most visible example comes from the investigation and subsequent lawsuit against Saskatchewan farmer Percy Schmeiser.

Schmeiser's farm near Bruno, Saskatchewan (a small town east of Saskatoon), has grown various strains of canola for more than thirty years. As a successful farmer, and former local politician, Schmeiser is well known in his community. Like many farmers prior to the introduction of herbicide-tolerant canola, he practised seed saving. He saved seed from particularly productive and healthy plots of land for subsequent replanting. Farmers who grow canola over many years usually select seeds from plants with larger pods that are more pilled and have a lower rate of disease.[17] The practice of seed saving and selection often produces a superior strain.

In 1997, Schmeiser noticed that one of his fields of canola was somehow different. This particular field followed a main power line where poles and ditches outlined the field. It is common practice for farmers to spray a herbicide such as Roundup (or generic versions based on the chemical glyphosate) to kill weeds and "volunteer" canola plants.[18] After spraying between the power poles and ditches, Schmeiser noticed that not all canola plants died. A subsequent respraying did not kill these plants either. In fact, after a few weeks, the density of canola plants increased, especially closer to the road.[19] Clearly, these plants were tolerant to the herbicide Roundup. This would prove to be a serious problem for Schmeiser since he never signed a Technology Use Agreement with Monsanto Canada to use its seed.

In July of 1998, Schmeiser received a telephone call from a representative of Monsanto Canada requesting that he allow Monsanto to inspect his land for the presence of its patented Roundup Ready canola. Schmeiser denied that he was using this patented canola and refused to permit inspection of his land. In August, he was served with a notice that Monsanto was suing him for infringing on its patent. In March 2001, he was found guilty of patent infringement. An appeal to the Supreme Court of Canada was launched, and in May 2004 the court upheld Monsanto's right to its patent.[20] How did Monsanto know that its patented canola could be found on this farmer's land?

After receiving regulatory approval from the Plant Products Division of Agriculture and Agri-Food Canada in 1995, Monsanto began selling in 1996 Roundup Ready canola seed to farmers. Many farmers participated in meetings with representatives from seed companies to learn more about this new product. They were informed about the attributes of this genetically modified seed and the nature of the Technology Use Agreement (TUA) that they would have to sign in order to use this product.[21] One of the conditions of the TUA was that Monsanto could inspect farms at any time for violations of the agreement. The tradition of seed saving was strictly forbidden by the TUA. Farmers would have to buy new seed every year and would be punished with heavy fines for violating this part of the agreement. In exchange for using this genetically modified seed, farmers would have to sacrifice some privacy and autonomy.

By requiring that farmers sign a TUA, Monsanto ensured a steady rate of return on its investment in this technology while simultaneously creating a culture of compliance among farmers. Since all canola farmers in a region paid the same amount per acre for the use of this seed, the profits associated with growing canola were contingent on complying with the TUA and ensuring that a "free rider" problem did not exist. In this case, a farmer who used this technology without paying the licensing fee and following the guidelines of the TUA would profit from the technology disproportionately at the expense of other farmers who were complying. How a community deals with free riders may have a significant impact on social cohesion.

It is very likely that canola farmers around Bruno, Saskatchewan, are aware of who in their community grows Roundup Ready canola. Meetings with seed company representatives, combined with informal discussions among farmers, help to identify users of this technology. Additionally, the open rural landscape in Saskatchewan allows individuals to see fields while passing by. To a trained eye, a dense collection of volunteer canola along the road or near hydro poles would suggest that something is different about these plants. Farmers who practise good land management would not allow such plants to grow too densely. A farmer suspected by fellow farmers of growing a patented crop without a licence may be ostracized socially and/or investigated.[22] This was likely the fate of Percy Schmeiser. A new dynamic in farming communities is being created by this culture of surveillance. Such a culture is low in trust and probably less stable and socially cohesive.

Does Agricultural Biotechnology Deskill Farmers?

In western Canada, farming has evolved over the years to become a highly productive system for generating food. In particular, farmers who grow grains and oilseeds require a high degree of skill to buffer the many market and nonmarket stresses that develop. For example, Canadian wheat farmers have been under tremendous pressure to cut prices in recent years. The North Dakota Wheat Commission has filed several complaints against the Canadian Wheat Board on behalf of the state's wheat producers alleging anticompetitive policies and practices.[23] American wheat farmers believe that the Canadian Wheat Board is systematically undercutting US competitors in the lucrative durum wheat export market by routinely underbidding US farmers by five to ten dollars US per bushel and by offering buyers additional protein content at no extra charge to secure contracts (McKenna 2001). In 2003, Canadian wheat farmers were levied a tariff of 8 percent on the export of wheat to the United States (McKenna 2003). Nonmarket stresses, like those accompanying climate change, require risk management skills and therefore create further challenges to farmers. In Saskatoon, the air was 30 percent drier in 2001 than in any other year since the 1890s (Mitchell 2001). Clearly, agriculture is a complex network of relationships that requires certain skills if farmers are to be successful.

One of the skills acquired by successful farmers in both the developed and the developing world is seed saving. As the example of canola in the previous section illustrates, seed saving allows farmers the option of replanting seeds from previous generations that have desirable agronomic traits. Much of the diversity in food that we presently enjoy comes from the practice of seed selection (Pueppke 2001). Broccoli, cauliflower, cabbage, and Brussels sprouts all come from the selective efforts of farmers over the years on a single species of plant (Bailey and Bailey 1976). Historically, as far back as 700 BC, artificial hybridization was practised by the Babylonians and Assyrians (Poehlman and Sleper 1995). The creation of new plants through hybridization increases biological (and dietary) diversity and diversifies farming. Unlike systems in which all farmers grow the same crop, a diversified system of farming creates niche markets with premiums. This helps to ensure the safety of the food supply by minimizing the risks associated with losses from biotic (e.g., pests) and abiotic (e.g., weather) pressures. Farmers in Canada have the option of growing a wide range of crops that best suit the market, soil, and weather conditions and their own skill sets. However, some of this choice is eroding due to the prohibition on seed saving for farmers who elect to grow certain genetically modified plants.

Farmers who grow Monsanto's Roundup Ready products (e.g., canola) are required by the company to sign a Technology Use Agreement that prohibits the practice of seed saving. For Monsanto, this clause ensures that farmers who grow its protected products continue to buy new seed from the company on a season-by-season basis. This represents a direct example of farmer deskilling. Farmers who grow these crops no longer have the option of selecting seeds from plants that grow best on their land. Most of the control that farmers had over their farming practices has been given away to companies such as Monsanto. In exchange for better weed control, farmers have sacrificed much. Their long-term financial security is more vulnerable to changes in the prices of purchased seeds and the herbicide with which they are designed to work. Many farmers probably welcome these changes since they remove some of the daily chores on the farm. Instead of worrying about seed selection and other related issues, farmers can spend more time cultivating larger tracts of land. Roundup Ready crops are easier for farmers to grow than nongenetically modified crops because they require less intensive management of the land (e.g., low tillage) and less work to control weeds. With historically low commodity prices, larger-scale operations are needed to increase profitability. A move toward larger farms is having profound impacts on rural communities across Canada.

Saskatchewan is known domestically and internationally as an agricultural province. With close to fifty million acres (or approximately twenty million hectares) of cultivated land, Saskatchewan has 65,995 farms (1996 Census of Agriculture data, Statistics Canada) with an average size of 1,152 acres (Saskatchewan Agriculture, Food and Rural Revitalization 2004). If we look at Canada as a whole, it becomes apparent that between 1976 and 1996 several changes in the relative proportion of differently sized farms occurred (Statistics Canada 1996). Between 1976 and 1996, the number of farms in Canada between 10 and 1,119 acres decreased on average by 25.5 percent. In contrast, farms less than ten acres increased by an average of 18 percent, while the largest farms (1,600 acres or more) increased by 33.4 percent. These trends suggest that small-scale (e.g., hobby farms) and large-scale farms are increasing in Canada while other farms are decreasing in number. What impact could this trend have on the social cohesion of rural communities? Are genetically modified crops magnifying these effects?

The erosion of the family farm is creating a range of social problems in rural Canada. Larger farms represent a greater concentration of wealth. Large farms are more heavily mechanized and require less units of labour on a per acre basis. As a result, rural communities become increasingly less able to provide job opportunities for many people. Many of today's young people leave rural communities for "greener pastures" in cities. Between 1971 and 1996, the percentage of people over sixty-five years of age living on farms in Canada increased from 6.1 to 8.3 percent (Statistics Canada 2000). The percentage of children under fifteen years of age decreased from 29.7 percent in 1971 to 20.6 percent in 1996. Although these trends mirror general demographic changes found in many parts of the developed world, in rural communities they tend to be more damaging.

Genetically modified crops introduce a new set of efficiencies into this equation. Since larger-scale farmers benefit more from the use of genetically modified crops (Altieri 2001), it is likely that since 1996 the number of medium-sized farms in Canada has continued to decline. The so-called Green Revolution of the 1950s had a similar effect. With the introduction of high-yield crops that required intensive applications of fertilizers and pesticides, many small farms disappeared around the world. Wealthier farmers with larger tracts of land benefited the most from the Green Revolution (Lappe et al. 1998). For poorer farmers, access to the new breed of fertilizers and pesticides proved to be difficult due to their relatively high costs. Nowadays, farmers who wish to grow genetically modified crops must pay a premium price for the privilege. Conflicts over patent rights and ever-increasing costs associated with scientific research are likely to make access to genetically modified seeds more expensive and exclusionary. In short, due to a range of pressures, rural communities are undergoing significant social transformations that affect social cohesion.

How Might Some Innovations in Agricultural Biotechnology Generate Conflict? The Case of Bt Corn and Refugia

Since the dawn of agriculture, farmers have been battling against insects intent on consuming their crops. In the early part of the twentieth century, a novel bacterium was discovered. This bacterium, named *Bacillus thuringiensis* (Bt), was effective at controlling larvae of the Mediterranean flour moth (Pueppke 2001). By 1927, Bt was commercially available as an insecticide (Feitelson 1993). Gardeners and organic farmers quickly adopted Bt as an effective and safe way to control certain insect pests. When consumed by moths and other insects, the Bt toxin produces a crystalline protein that paralyzes their gut muscles, resulting in starvation (Yamamoto and Powell 1993).

Advances in recombinant DNA technology in the 1980s led to the development of plants that could express Bt protein. The gene responsible for producing this protein in the bacterium was incorporated into the genome of certain plants (e.g., early experiments concentrated on tomatoes) so that the insecticide would be produced internally by the plant instead of being applied topically. Results of this early work were promising. Valuable crops such as corn and cotton could be protected against the European core borer and cotton boll weevil by inserting the gene for Bt production into the plants themselves. Although Bt crops are not widely grown in western Canada, the problems that arise from them are emblematic of how advances in biotechnology affect different kinds of communities.

Like many developments in biotechnology, the creation of plants that express Bt toxin in their tissues generates a host of scientific and nonscientific issues. The scientific issues that arise over the use of these transgenic plants include concerns over possible harm to nontarget insects such as the monarch butterfly (Losey, Rayor, and Carter 1999), impacts on biodiversity due to the biological activity of the Bt toxin in soil (Saxena, Flores, and Stotzky 1999), and concerns about the safety for humans who consume plants modified in this fashion.

Scientists and farmers alike are also very concerned about a management issue arising from the use of this technology. As with other kinds of insecticides, the development of resistant colonies of insects is a serious threat. For example, in a population of European corn borers, a very small percentage of insects will be completely resistant to Bt toxin. Normally, this resistance presents little concern since these insects are most likely to mate with nonresistant insects (due to their small numbers), thereby producing progeny that are susceptible to the high concentrations of Bt found in fields of transgenic corn or cotton. However, if the use of these protected plants were widespread enough, only insects with the gene for resistance to Bt would survive. Eventually, a colony of resistant insects would emerge due to the mating of resistant with resistant insects. Bt (both the transgenic and the topical varieties used by organic farmers) would become virtually useless as an insecticide in this instance.

To reduce the risks associated with the formation of resistance, farmers are being asked voluntarily to build what are called refugia into fields. Refugia are zones that encourage the reproductive success of the insect of concern. Plots of nongenetically modified plants are placed strategically within fields of genetically modified plants so that a colony of nonresistant insects can survive. Farmers should not spray refugia with any insecticides. In theory, refugia prevent the development of a resistant colony by providing the few resistant insects with an opportunity to mate with a larger number of nonresistant insects. It is assumed that such practices will dilute the prevalence of Bt-resistant insects, thereby ensuring the continued successful use of Bt. Companies such as Syngenta recommend to farmers that 20 percent of their cornfields should be devoted to non-Bt refugia (Syngenta 2001).

The proper use of refugia by farmers who grow Bt crops is important for other reasons too. Farmers may be reluctant to "sacrifice" 20 percent of their fields to refugia. Improper practices such as poorly configured or placed refugia may also nullify the benefits and create conflicts with neighbouring farmers. Since fields are usually contiguous, a certain degree of cooperation is required between farmers who plant the same kinds of crops. A farmer who plants a field of Bt corn without an adequate amount of refugia jeopardizes the security of his neighbours who grow Bt, conventional, and organic corn. A colony of resistant insects will move between fields, with little concern for private property rights, and damage crops. To combat a colony of resistant insects, farmers growing Bt corn will need to use chemical insecticides. They increase the cost of doing business and generate possible environmental and human health risks. Organic farmers are more seriously affected. For many, the use of sprayed Bt is an effective strategy for combatting insect pests. Although other techniques for controlling pests are available to organic farmers, Bt is probably the best tool available. A spreading colony of Bt-resistant insects could devastate organic farmers. It is likely that farmers who grow Bt crops will be under increasing scrutiny from other farmers interested in their field management practices. Future lawsuits might even emerge from farmers who believe that poor management practices of neighbours have affected their livelihoods. Such issues were less serious prior to the advent of transgenic plants capable of expressing the Bt toxin. Again the social cohesion of agricultural communities is exposed to a different kind of stress resulting from the introduction of a commercially valuable insect-control technology such as Bt corn.

Conclusion

By examining how the introduction of genetically modified crops such as herbicide-tolerant canola and Bt corn affects farming communities in western Canada, this chapter has demonstrated how innovations in agricultural biotechnology affect social cohesion. These innovations benefit more large-scale farmers, lead to a culture of surveillance, deskill farmers, and generate conflict. Based on the evidence presented, the conclusion seems to me to be inescapable: innovations in agricultural biotechnology are leading to, or at least coinciding with, a decline in the social cohesion of agricultural communities. What are the implications of this conclusion?

If highly cohesive communities are tightly coupled and robust and share common interests and goals, then weakly cohesive communities are diffuse, unstable, and divisive. The latter communities are more likely to suffer when economic fortunes decline and more likely to tolerate disparities in wealth, income, and power. Such communities are less able to muster and nurture the "social capital" needed to sustain themselves. As a result, weakly cohesive agricultural communities represent a decline in the quality of life for individuals living in rural parts of western Canada. Some may argue that this decline began many years before the introduction of agricultural biotechnology. However, it is worth noting that herbicide-tolerant canola and Bt corn were the innovations of an agricultural system in which international actors such as Monsanto benefit from weakly cohesive agricultural communities. Although such companies are not solely responsible for the changes that are occurring in rural communities, the introduction of agricultural innovations that stress agronomic benefits has many unexpected effects. One of these effects is a weakening of social cohesion in these communities. Clearly, much more research is needed in this area.

Notes

1 A "conventional farmer" uses chemical inputs that are often derived from petroleum products. These inputs include a wide range of pesticides, herbicides, and fertilizers.

2 See Chapter 3 for a breakdown of the genetically modified crops planted globally.

3 In the spring of 2001, a genetically modified type of flax named Triffid was deregistered. Reportedly, concerns about the reluctance of export markets to accept GM products led to Canadian flax producers losing interest in growing this crop.

4 In 2000, Canadian farmers planted 12.1 million acres (or approximately 4.9 million hectares) of canola. Almost 80 percent of growers used at least one herbicide-tolerant system. Close to 50 percent grow transgenic canola (Koch Paul Associates 2001).

5 Canola has been modified to increase the amount of linoleic fatty acid (omega –6) and alpha-linolenic fatty acid (omega –3). Both of these fatty acids are associated with a healthy diet. Canola oil is widely recognized as containing one of the lowest levels of saturated fatty acid of any vegetable oil.

6 A study by Koch Paul Associates in 2001 suggests that Roundup Ready canola accounts for almost 71 percent of transgenic canola grown. Liberty Link accounts for the remainder.

7 Roundup is the best-selling agricultural chemical on the market. In 2000, Monsanto sold US$2.8 billion worth of Roundup. Roundup sales, which account for half of Monsanto's revenue, are why Monsanto reported a solid second quarter in 2001 (Barboza 2001).

8 Quotation from Aventis's website entitled "LibertyLink for Canola." On-line at <http://www.libertylink.ca/index.shtml>, retrieved 3 August 2001.

9 Biotechnology companies such as Monsanto do not sell their seed, genetically modified or otherwise, directly to farmers. In contrast, Aventis does.

10 Concerns range from pollen drift and the potential to create "superweeds," the problem of herbicide tolerance, and the safety of consuming this kind of genetically modified crop. In the latter case, the oilseed industry is quick to point out that "properly" processed oil from plants such as canola is unlikely to contain any protein or DNA. In other words, the processed oil from herbicide-tolerant canola is indistinguishable from the oil derived from non-GM canola.

11 Technology Use Agreements (TUAs) are contracts that farmers sign with some seed companies to buy seed at a fixed price and to purchase particular quantities of a selective herbicide. Such agreements often give the parent company (e.g., Monsanto Canada) the right to inspect a farmer's land for compliance and levy stiff penalties in the event of noncompliance.

12 Until just recently, the average consumer was probably unaware that genetically modified foods were already present in many processed foods.

13 On 10 January 2002, a group of organic farmers in Saskatchewan filed a class-action lawsuit against Monsanto and Aventis for damage caused by genetically modified canola. This group is also interested in stopping Monsanto from introducing a herbicide-tolerant strain of wheat. The Saskatchewan Organic Directorate claims that "our ability to farm organically is being threatened" (Wong 2004). In May 2004, Monsanto agreed to pull from the regulatory approval process, for the time being, its variety of GM wheat.

14 Another example of how farming communities are changing is the creation of "bedroom" communities near major cities. Villages and towns that once supported local farming communities are becoming populated by people leaving the cities.

15 Traditional polyculture involves growing more than one kind of crop and taking full advantage of crop rotation and different root depths or soil nutrients.

16 This assertion is by no means universally true. For example, some small farmers who raise cattle or swine cannot afford state-of-the-art equipment for dealing with farm waste. The E. coli found in the municipal drinking water supply of towns such as Walkerton, Ontario, was due to fecal material from farms.
17 Canola is an oil crop. Its seeds are crushed and made into oil. The seeds are found in pods that look somewhat similar to pea pods. Larger pods with a higher density of seeds produce more oil.
18 "Volunteer" refers to canola plants that are undesirable because of either where they grow or traits that they may possess. Farmers usually destroy these plants with a nonselective herbicide so as to prevent them from reproducing. Such herbicides are also used to "preburn" fields prior to the planting season to reduce weed growth.
19 Schmeiser would eventually claim that seed from untarped trucks had contaminated his land. Since canola is a light seed, it is possible that some contamination occurred. Many farmers near the Schmeiser farm were growing Roundup Ready canola and would use this particular road to transport their goods.
20 The court's decision can be found on-line at <http://www.lexum.umontreal.ca/csc-scc/en/rec/html/2004scc034.wpd.html>.
21 At the Examination for Discovery in the *Monsanto v. Schmeiser* case, it was revealed that farmers were not warned about the possibility that genetically modified canola seeds could spread to adjoining lands; nor were they told to warn their neighbours about this possibility. In fact, Monsanto does not advise or require that farmers maintain buffer zones or that they place tarps over vehicles carrying seed.
22 Monsanto Canada maintains a toll-free telephone number for receiving such tips. Anecdotal accounts suggest that farmers who provide such tips receive a gift from Monsanto (e.g., a jacket).
23 To learn more about the specific concerns of the North Dakota Wheat Commission, visit its website at <http://www.ndwheat.com>.

References

Altieri, M.A. 2001. "Genetically Engineered Crops: Separating the Myths from the Reality." *Bulletin of Science, Technology, and Society* 21: 130-46.

Bailey, L.H., and E.Z. Bailey. 1976. *Hortus Third.* New York: Macmillan.

Barboza, D. 2001. "Monsanto Booms – But Is Heavily Dependent on Roundup." *New York Times,* 2 August.

Buttel, F., O. Larson, and G. Gillespie Jr. 1990. *The Sociology of Agriculture.* Westport, CT: Greenwood Press.

Durkheim, É. 1893. *The Division of Labor in Society.* New York: Simon and Schuster.

Feitelson, J.S. 1993. "The *Bacillus Thuringiensis* Family Tree." In L. Kim, ed., *Advanced Engineered Pesticides,* 63-71. New York: Marcel Dekker.

Fliegel, F. 1993. *Diffusion Research in Rural Sociology: The Record and Prospects for the Future.* Westport, CT: Greenwood Press.

Koch Paul Associates. 2001. *An Agronomic and Economic Assessment of Transgenic Canola.* Prepared for the Canola Council of Canada. On-line at <http://www.canola-council.org/production/gmo_toc/html>.

Lappe, F.M., J. Collins, P. Rosset, and L. Esparza. 1998. *World Hunger: Twelve Myths.* 2nd ed. New York: Grove Press and Earthscan.

Leopold, A. 1949. *The Sand County Almanac.* New York: Ballantine.

Losey, J.J.E., L.S. Rayor, and M.E. Carter. 1999. "Transgenic Pollen Harms Monarch Larvae." *Nature* 399: 214.

McKenna, B. 2001. "Wheat Board Policies Seem Unfair to U.S." *Globe and Mail,* on-line edition, 22 December.

–. 2003. "Canadians Furious with U.S. Wheat Tariff." CBC News, on-line at <http://www.cbc.ca/stories/2003/05/02/wheat_030502>, 2 May.

Mitchell, A. 2001. "Canada's Weather Bizarre in 2001." *Globe and Mail,* on-line edition, 27 December.

Poehlman, M.P., and D.A. Sleper. 1995. *Breeding Field Crops*. 4th ed. Ames: Iowa State University Press.

Pueppke, S. 2001. "Agricultural Biotechnology and Plant Improvement." *American Behavioral Scientist* 44: 1233-45.

Rogers, E. 1962. *Diffusion of Innovations*. New York: Free Press of Glencoe.

Saskatchewan Agriculture, Food and Rural Revitalization. 2004. "Crops." On-line at <http://www.agr.gov.sk.ca/Crops.asp?firstPick=Crops>.

Saxena, D., S. Flores, and G. Stotzky. 1999. "Transgenic Plants: Insecticidal Toxin in Root Exudates from Bt Corn." *Nature* 402: 480.

Statistics Canada. 1996. Census of Agriculture, "Farms by Size of Farms: Canada, 1976-1996." Ottawa: Statistics Canada.

–. 2000. "Farming Facts, 1999." Canada: Minister of Industry. On-line at <http://www.statcan.ca/english/freepub/21-522-XIE/9900121-522-XIE.pdf>.

Syngenta. 2001. "Commit to Planting a Refugia on Your Farm." On-line at <http://www.nkcanada.com>, posted 5 November.

Wong, C. 2004. "Saskatchewan Organic Farmers File Lawsuit against Monsanto and Aventis." On-line at <http://www.mindfully.org/Farm/2004/Monsanto-Saskatchewan-Organic10jan04.htm>.

Yamamoto, T., and G.K. Powell. 1993. "*Bacillus Thuringiensis* Crystal Proteins: Recent Advances in Understanding Insecticidal Activity." In L. Kim, ed., *Advanced Engineering Pesticides*, 3-42. New York: Marcel Dekker.

3

Agricultural Biotechnology and Developing Countries: Issues of Poverty Alleviation, Food Security, and Sustainable Development

Jacqueline E.W. Broerse and Joske F.G. Bunders

Biotechnology is often presented as a potentially powerful factor in contributing to poverty alleviation, food security, and sustainable development in developing countries. Looking at the innovations currently developed through biotechnology research and development (R&D), we can conclude, however, that these innovations are usually inappropriate for these purposes. Skepticism therefore prevails in the development community about the usefulness of biotechnology as an instrument in poverty alleviation, food security, and sustainable development. In this chapter, we provide a broad overview of a range of recent developments in agricultural biotechnology, primarily, but not exclusively, focusing on the developing world. We then proceed with a more in-depth discussion of the threats and opportunities of biotechnology for small-scale agriculture. It is then argued that, if biotechnology is to benefit the poor, a double shift in the research paradigm is needed – biotechnology R&D should be specifically (1) focused on agro-ecological systems and products important to poor people and (2) contextualized within the broader socioeconomic and cultural situations of the poor and with a deeper understanding of sustainability issues. The challenge of institutionalization of interactive and participatory approaches to existing organizations is also discussed. The chapter closes with an examination of future perspectives of biotechnology for the poor.

Biotechnology research and development are conducted in a world more than ever in need of innovative products. Alleviation of poverty, eradication of hunger, and enhancement of sustainable development in developing countries present formidable tasks for the coming decades. The World Bank (1997) has estimated that over 1.3 billion people in developing countries are profoundly poor, having incomes of one dollar US per day or less per person, and that a further three billion people live on less than two dollars US per day. It is further estimated that about 800 million people, particularly located in Asian and African countries, do not have access to sufficient food to meet their needs, despite increases in overall food availability and declining food prices (Pinstrup-Anderson and Pandya-Lorch 2000). The International Food Policy Research Institute (IFPRI 1997) has projected that world grain production will need to increase by 40 percent between 1995 and 2020 to keep pace with population growth. Since further expansion of the area cultivated is not feasible in most of the world's developing regions, increases in food production will have to be achieved by a more efficient use of land already under cultivation (Bunders, Haverkort, and Hiemstra 1996; ECA 2002; Serageldin 2000; van Wijk 2000). Technological innovation, generated through scientific research, will be vital in bringing about such changes.

Biotechnology is currently one of the most challenging areas of technological development. In principle, it offers many promises for the world community given its potential to address problems not resolved by conventional research, to speed up research processes, and to increase research precision. For example, it could contribute to improved human health by offering new ways to understand the genetic basis of diseases, to improve diagnostics, and to develop drugs and vaccines for treatment of diseases (Serageldin and Persley 2000). In agriculture and forestry, it holds the promise of improving the biological potential of crops, livestock, fish, and trees, reducing the need for agrochemicals, adapting plants to harsh growing conditions (such as drought, salinity, and pests), and improving health by introducing desirable nutritional characteristics in food crops (Braunschweig 2000; Cohen 1999; ECA 2002; James 2002; Serageldin 2000).

Life science companies claim to be prepared to take up the challenge of feeding the world through biotechnology, as indicated on the websites of companies such as Monsanto, DuPont, and Novartis (see also van Wijk 2000) and in advertisements of Monsanto in nation-wide newspapers in the United Kingdom and the United States with the slogan "Food * Health * Hope" (see also Sexton, Hildyard, and Lohmann 1998). Given their dominant positions in biotechnology research, these companies seem, at first glance, to be well placed to fulfill this promise of biotechnology. But Anatole Krattiger (2000) questions whether their deeds will really live up to their words. There is also a very familiar ring to these claims. Similar claims were made in the early 1980s, only then by biotechnologists instead of life science companies, and hardly any of them have been substantiated today.

To further temper the optimism, Ismail Serageldin (2000), chair of the Consultative Group on International Agricultural Research (CGIAR) and vice president of the World Bank for Special Programs, has warned that the potential benefits of biotechnology should not divert attention from the real concerns about the application of the new science. The perils of biotechnology lie in the ethical issues concerning genetic modification and the risks to human health and the environment associated with the production and consumption of genetically modified organisms (GMOs) and foods. These issues have led to rising public concerns about various applications of modern biotechnology, particularly in Europe (e.g., Juma 2000).

Other perils relate to the possibility that certain applications could further increase the prosperity gap within and between developing countries and between industrial and developing countries. According to Klaus Leisinger (1999), this latter gap may grow because biotechnology introduces functionally identical, novel substitutes for traditional raw materials (e.g., through genetically modified enzymes) and permits the production of existing raw materials in industrial rather than agricultural settings (e.g., through plant tissue culture). Other concerns relate to the impact of the Genetic Use Restriction Technologies (GURTs), such as the terminator gene, which enhance the benefits from innovations in seed development – farmers purchasing the seed are not able to reproduce it (either for sale or own use) (ECA 2002; Swanson and Goschl 2000). Moreover, there are fears that the industrial world may not adequately compensate the developing world for exploiting its indigenous knowledge and genetic resources while selling back the resulting products to developing countries at high prices (Biber-Klemm 2000; Dutfield 2000; ECA 2002).

Furthermore, there is more to poverty alleviation, food security, and sustainable development than "simply" developing crop varieties with improved traits through biotechnology (ECA 2002; Serageldin 2000; van Wijk 2000). Generating technological innovations that are appropriate in the diverging ecological and socioeconomic contexts of small-scale agriculture – the predominant form of agriculture in developing countries – has proved to be quite complicated (Broerse 1998; Bunders, Haverkort, and Hiemstra 1996; van Wijk 2000; see Mehta, this volume, Chapter 2).

In the next section, we highlight general trends in biotechnology research and application in order to provide a broader worldwide picture before discussing various issues in developing countries.

Worldwide Trends

Modern biotechnology was started in the early 1970s and was mainly conducted for the next two decades by publicly funded research institutes and universities. During the 1990s, this picture changed drastically. Nowadays, the private sector is the driving force in biotechnology R&D. With slightly decreasing public sector investments in R&D, the private sector currently contributes approximately 80 percent of all R&D investments. Ismail Serageldin and Gabrielle Persley (2000) have estimated that this 80 percent is coming mainly from six large companies (previously, many more small companies were involved).

Due to the dominance of the private sector – a position protected by intellectual property rights (notably patents)[1] – the products that are developed are oriented toward large and lucrative markets, such as diagnostics (immuno-assays and DNA probes), human pharmaceutical and animal vaccines, plant improvements (addition of single gene traits such as herbicide or pest resistance), and food processing (effluent treatment, "natural" additive production, and enzyme-based processing). The majority of biotechnology products are applied in the health care sector.

In the mid-1990s, the first generation of genetically modified (GM) crops was grown on a large scale. Each year since 1996, Clive James has prepared a report on the global status of commercialized transgenic crops, published by the International Service for the Acquisition of Agri-Biotech Applications (ISAAA). James has estimated that in 2002 approximately sixty million hectares of land were planted worldwide with transgenic varieties of over twenty plant species by almost six million farmers. During the past seven years, the increase in area has been well over 10 percent per year. The adoption pattern is, however, highly skewed. So far, only four countries account for 99 percent of the total area planted with GM crops: the United States (66 percent), followed by Argentina (23 percent), Canada (6 percent), and China (4 percent of global GM area in 2002). Only twelve other countries were growing GM crops commercially in 2002: seven developing countries (South Africa, Mexico, Indonesia, India, Colombia, Uruguay, and Honduras), three industrial countries (Australia, Spain, and Germany), and two Eastern European countries (Romania and Bulgaria) (James 2002). Commercially, the most important transgenic plant species were soybean (62 percent), maize (21 percent), cotton (12 percent), and canola (5 percent of global GM area in 2002). James has also investigated the traits included in most developed GM crops: the dominant trait is herbicide tolerance (soybean, cotton, maize), followed by insect resistance (cotton, maize).

The main benefits of these initial varieties are better weed and insect control, higher productivity, and more flexible crop management. The societal groups that have so far primarily benefited from these GM crops are agribusinesses and farmers in intensive agricultural systems (by reducing yield variability and unit production costs; Pingali and Traxler 2002). Most of the GM crops were destined for the animal feed market.

Other crop-input trait combinations being field-tested include virus-resistant melon, papaya, potato, squash, tomato, and sweet potato; insect-resistant rice, soybean, and tomato; disease-resistant potato; and delayed-ripening chili pepper. Research is also aimed at modifying oil content (rapeseed, canola), increasing the amount and quality of protein (maize), or increasing vitamin content (rice) (James and Krattiger 1999). Much greater emphasis is now being put on improving the nutritional value of food crops, thereby extending the benefits of GM crops to consumers. Interesting examples include new rice varieties that contain beta carotene, a precursor to vitamin A (so-called golden rice; Dawe, Robertson, and Unnevehr 2002), essential amino acids, and iron (ECA 2002). There is also work in progress to use plants, such as maize, potato, and banana, as biofactories for the production of vaccines and high-value human proteins (mainly therapeutics) (Serageldin and Persley 2000). This latter use is known as molecular farming. Researchers in many countries are looking at using oil crops such as canola to produce medically valuable proteins such as Hirudin (an anticoagulant normally derived from leeches) to improve the quality and increase the quantity of biologic agents.

Developments in livestock production are less spectacular. Modern biotechnology is mainly applied to livestock in the field of breeding and health, particularly speeding up the reproductive process, enabling efficient selection of improved breeds, and preventing diseases (vaccines). Various animal genome projects are undertaken to shorten the discovery process of potentially interesting genes that could improve livestock. However, the relatively high costs and low efficiencies of animal breeding programs compared to plant breeding favour investments in the latter (Serageldin and Persley 2000). The genetic modification of animals is currently mainly focused on the production of high-value human proteins.

The Developing World

What is happening in developing countries? In answering this question, one thing is clear: it is impossible to talk about *the* developing world. The differences are enormous. In this section, we illustrate this diversity in the following fields: ag-biotech policies, state-of-the-art biotechnology, funding of biotechnology R&D, biosafety, intellectual property rights, and public acceptance. Also discussed is the role of the international development community.

Ag-Biotech Policies

Agriculture is still the largest economic sector in terms of income, employment, and foreign exchange earnings in many developing countries (Falconi 1999). Recognizing the strategic importance of biotechnology, various governments have formulated ag-biotech policies – some of them substantial and some only exploratory – for example in Brazil, Burundi, China, Colombia, Costa Rica, Egypt, Honduras, India, Indonesia, Ivory Coast, Jordan, Kenya, Malaysia, Mexico, Philippines, Singapore, South Africa, Thailand, Vietnam, and Zimbabwe (Serageldin 2000). These policies usually concern capacity building in research and development and involve research and training programs in ag-biotech, the establishment or assignment of a specific centre (or centres) of excellence on biotechnology research, and some (modest) resource allocations. Very few countries, however, have set specific research priorities (Braunschweig 2000) or policies that may enable them to make use of private-sector research spillovers in particular and that stimulate public-private collaborations in general (Bustamante and Bowra 2002; Pingali and Traxler 2002).

State-of-the-Art Biotechnology

There is considerable work in progress on the use of modern biotechnology in developing countries (see Figure 3.1), but there are large differences between developing countries in R&D capacity and investment. Particularly, the emerging economies (including Argentina, Brazil, Mexico, China, India, Thailand, and South Africa) are making major investments in biotechnology, especially in crop breeding, as is witnessed by the progress made in biotechnology research in these countries (Serageldin and Persley 2000). For example, in 1993, China was the first country in the world to grow GM crops on a commercial scale – virus-resistant tobacco and tomato (Skerritt 2000). To date, over fifty genetically modified organisms have already been approved for small-scale field testing, environmental release, or commercial production. Various traits were targeted for improvement, including disease, insect, and herbicide resistance, and quality improvement in crops such as rice, wheat, corn, cotton, tomato, pepper, potato, cucumber, papaya, and tobacco (Zhang 2000). In Mexico, field testing started in 1988, with trials of plants with genetically improved insect resistance (maize and cotton), herbicide and virus resistance (potato), and delayed ripening (tomato) (Alvarez-Morales 2000). Some of these crops, such as the virus-resistant potato, are now being grown by Mexican farmers. In 1999, over 140 permits were granted for the release of genetically modified organisms. It should be noted, however, that part of the field trials conducted in developing countries concerns new crop varieties developed by life science companies wishing to introduce their materials into the markets of these countries. For example, in Thailand, the Flavr Savr™ tomato produced by Calgene for the generation of seeds was field-tested in 1994, a field trial of Monsanto Bt cotton was carried out in a netted house in 1996, and in 1997 a Bt cornfield trial was carried out in a netted house by Novartis (Tanticharoen 2000).

However, in most countries in sub-Saharan Africa, developments in genetic modification are in their infancy, although many countries have a reasonable capacity in the less advanced techniques, such as tissue culture and genetic markers. Ag-biotech research is particularly focused on crops such as maize, cassava, sorghum, banana, and bean (in order of importance), with a strong emphasis on developing plant resistance to insect pests and viral, bacterial, or fungal diseases (Komen, Mignouna, and Webber 2000). In this part of the world, few field trials of GM crops have taken place. A

Figure 3.1

Examples of products ready for diffusion in developing countries

- *Disease-free planting material:* Various tissue-culture techniques are applied for the micro-propagation of disease-free planting material. They involve mainly export crops, such as coffee, banana, oil palm, and sugar cane.
- *Biocontrol agents:* Products include biopesticides based on *Bacillus thuringiensis* and *B. sphaericus* and a pheromone-based attractant decoy for tick vector control.
- *GM crop varieties:* Virus-resistant potato, tobacco, and tomato and insect-resistant cotton have been field-tested in various developing countries.
- *Diagnostics and vaccines for livestock diseases:* Diagnostic tests and recombinant DNA vaccines are used for rinderpest, cowdriosis (heartwater), theileriosis (East Coast fever), and foot-and-mouth disease.

Source: Cohen, Falconi, and Komen (1998).

notable exception is South Africa, where GM research has mainly focused on developing pest-resistant cotton, maize, and strawberry varieties. The first commercial releases of GM crops took place in 1997 (ECA 2002; Njobe-Mbuli 2000). Success stories from other countries in sub-Saharan Africa include the adoption by farmers of tissue-cultured, disease-free banana plantlets in Kenya (Qaim 2000) and the use of new vaccines against animal diseases in Kenya and Zimbabwe (Persley 2000). For most other countries in sub-Saharan Africa, the lack of adequate national research capacity has been identified as a major hurdle for the application of modern biotechnology (Braunschweig 2000).

Funding of Biotechnology R&D

Much of the R&D in developing countries is being funded by national governments (about two-thirds), while the remaining investments mainly come from bilateral and multilateral development agencies (Falconi 1999). However, there are large differences between developing countries. Particularly in many countries in sub-Saharan Africa, agricultural research is increasingly dependent on donor funding, often constituting almost half of agricultural research spending (Braunschweig 2000; Komen, Mignouna, and Webber 2000). However, financial assistance from these sources has also fallen (ECA 2002).

Private-sector participation has generally been quite limited in most developing countries. Only a few commercial companies are engaged in modern biotechnology research. Most private companies are specialized in tissue culture of fruits and ornamentals, which is less costly, less risky, and closer to the market (Falconi 1999). However, in some larger countries in Asia and Latin America, private-sector investments in applied research can be substantial, reaching almost 40 percent of total research expenditures (Braunschweig 2000).

Biosafety

As well as building their capacity in research, many developing countries have recognized the need to build capacity in the safe handling of GMOs (biosafety). During the 1990s, various countries took steps to develop biosafety regulations and have established a biosafety commission in order to reduce the human health and environmental risks associated with the release of GMOs. Cohen, Falconi, and Komen (1998) have estimated, however, that only about 10 percent of developing countries have actually established biosafety regulations/guidelines. These countries include Brazil, China, Egypt, India, Indonesia, Kenya, Mexico, the Philippines, South Africa, Thailand, and Zimbabwe.

Establishing guidelines is relatively easy, while establishing a well-functioning system for biosafety review is much more complex (ECA 2002; Juma 2000; Komen, Mignouna, and Webber 2000; Madkour, El Nawawy, and Traynor 2000). It should also include the availability of knowledge and infrastructure for (1) contained and large-scale field testing of GMOs, (2) inspection of approved activities in particular, and research activities in general, to ensure compliance with biosafety requirements, and (3) monitoring of environmental release of GMOs to collect biosafety data for improved risk assessment. Such an infrastructure, however, is rarely in place in developing countries. One notable exception is South Africa, which has set up, besides legislation, a regulatory watchdog agency, SAGENE (Rybicki 1999). Developing countries are increasingly pressured by industrialized countries to establish a biosafety system (Braunschweig 2000).

Intellectual Property Rights

In most developing countries, policies for the application and/or management of intellectual property rights (IPR) regarding biotechnological products are still under formulation. Many developing countries have traditionally been reluctant to enforce any form of IPR, particularly patents, because doing so would limit their access to foreign technologies (van Wijk, Cohen, and Komen 1993). The introduction and enforcement of IPR could thus adversely affect their domestic industries.

However, during the past decade, many developing countries are increasingly agreeing to accept international IPR standards because they may reduce the technology gap. The dominance of the private sector in biotechnology R&D makes developing countries more dependent on international collaborations; these collaborations may be the only way to gain access to protected technologies from abroad without being subjected to trade sanctions and suspension of technological cooperation (Pingali and Traxler 2002). Increasingly, research institutes in developing countries codevelop technologies with life science companies. In addition, the local private sector pressures for some form of IPR to protect its relatively high investments. The more outward oriented the national industrial and agricultural sectors, the more open the economy, and the greater the level of technological development, the more there is to be gained from stronger IPR legislation (van Wijk, Cohen, and Komen 1993).

Also in this case, however, it is one thing to have IPR legislation in place, yet another to manage it effectively. Recent surveys of the International Service for National Agricultural Research (ISNAR) (e.g., Cohen, Falconi, Komen, and Blakeney 1998; Salazar, Falconi, Komen, and Cohen 2000) have revealed that both CGIAR centres and national agricultural research organizations in developing countries currently lack institutional mechanisms and policies/guidelines to deal with matters related to IPR.[2] This is usually left to individual researchers – whether it concerns the patenting of innovations or the acquisition of patented technologies and constructs – who often do not have in-depth knowledge of the topic.

Role of the Public

The issue of public concern and consumer acceptance of certain applications of modern biotechnology has hardly been featured on policy agendas in most developing countries. Although few surveys on attitudes toward biotechnology are conducted in the developing world, in many developing countries there seems to be neither strong objection from the general public nor opposition from societal groups. Surveys in Asia indicate that public opinion in China, Thailand, and India is quite supportive (Macer, Azariah, and Srinives 2000).

Negative views are witnessed in a few Latin American countries. In Brazil, the approval of the market introduction of Monsanto's genetically modified soybean was postponed to at least mid-2000 – awaiting an appeal case in higher courts – due to heavy opposition from a Brazilian consumers' institute in collaboration with Greenpeace (Sampaio 2000). In Mexico, the introduction of genetically modified maize is heavily opposed by some societal groups (Alvarez-Morales 2000). Due to this opposition, the Mexican government is reconsidering its biotechnology policy, imposing severe restrictions on field trials with GM maize varieties, and investigating the establishment of an elaborate monitoring system for environmental GMO releases.

Nevertheless, even in countries where consumers do not oppose GM food, the issue is considered relevant because many agricultural products are exported to industrialized countries where public perception of GM crops is less favourable (Braunschweig 2000).

International Development Community

The international development community has been involved since the mid-1980s in enhancing biotechnology capacity in developing countries and supporting these countries in related policy issues. Donor organizations and agricultural research institutes recognized that the application of biotechnology to developing country agriculture posed specific challenges. In order to enhance the potential benefits and to minimize negative impacts, the international donor community undertook the following activities (Komen 1999):

1 Support of the establishment of (a) a specialized international institute for biotechnology, the International Centre for Genetic Engineering and Biotechnology (ICGEB), with locations in New Delhi (India) and Trieste (Italy), and (b) specialized national facilities for biotechnology in several developing countries.
2 Creation of special biotechnology initiatives and networks (e.g., the Cassava Biotechnology Network) aimed at transferring advanced technology to developing countries.
3 Development of biotechnology capacity in established international research programs, including the International Agricultural Research Centres (IARCs) of the CGIAR, which have become key players in agbiotech for national agricultural research centres in developing countries (Cohen, Falconi, Komen, and Blakeney 1998).
4 Strengthening of the international component and outreach activities of advanced national research institutes in industrialized countries.
5 Facilitating technology transfer from companies holding proprietary biotechnologies to noncommercial or otherwise segmented markets in developing countries. The International Service for the Acquisition of Agri-Biotech Applications (ISAAA) has been established with this specific aim.[3]
6 Support of special advisory programs on the policy and management aspects of ag-biotech: for instance, the Intermediary Biotechnology Service (IBS), established by CGIAR's ISNAR, and the Food and Agriculture Organization's (FAO) biotechnology program (Spillane 2000).

Lack of International Initiatives Focused on the Poor

From the above analysis, we can conclude that the biotechnology revolution is in full swing, but it has so far been very much the preserve of the richer countries and, to a lesser extent, some emerging economies. Most biotechnology R&D takes place in private- and public-sector laboratories in the developed world; the biotechnological R&D capability in Third World institutions is, with few exceptions, relatively limited. In the next section, we focus on poor producers in developing countries – small-scale, resource-poor farmers – in order to illustrate the opportunities and threats of biotechnology in more detail in a concrete setting.

Small-Scale Agriculture

The problems of small-scale, resource-poor farmers rarely feature on research agendas either in the industrialized countries or in the developing world. With their lack of economic power and their location in difficult growing environments with a poor infrastructure, these farmers are unlikely to exert any effective "demand pull" on private-sector research agendas (Pingali and Traxler 2002; Spillane 2000). The life science companies are first and foremost focusing on markets that allow them to recoup their expensive investments in research and development (Krattiger 2000; Pingali and Traxler 2002). These companies thus have little incentive to develop products specifically for small-scale, resource-poor farmers.

Due to a lack of political power, such farmers are equally unable to codetermine biotechnology R&D agendas of the public sector. As Charles Spillane (2000, 2) explains, "In most public sector institutions or funding bodies there are currently few priority setting or needs assessment mechanisms (analogous to the functions of private sector marketing department) in place to help guide the direction of publicly-funded plant biotechnology or crop improvement research towards meeting the immediate needs of poorer farmers or consumers." It is therefore likely that innovations created in and for industrialized countries will be transferred to the developing world. On the basis of the experiences with earlier efforts to introduce Western agricultural innovations in developing countries, such as the Green Revolution (see Figure 3.2), it can be expected that biotechnological innovations resulting from such a nonpoor focus will tend to be irrelevant or detrimental to many farmers, since these are mainly being developed in a similar centralized, top-down way as the Green Revolution technologies (Bunders and Broerse 1991; Manicad and McGuire 2000; Persley 2000).

Although small-scale farmers in developing countries have virtually no access to mainstream biotechnological R&D (being too expensive or in other ways inappropriate), they are nevertheless intimately affected by its application by others. The biotechnologies designed for intensive agriculture are usually first and foremost adopted by resource-rich farmers. These farmers are likely to capture most of the benefits through expanded production and reduced unit costs (Persley 2000). The assumed "trickle-down effect" from resource-rich to resource-poor farmers rarely occurs (van de Sande, Ruivenkamp, and Malo 1996). As a result, the gap in yield and income between poorer and richer farmers increases. Results of recent adoption studies of specific biotechnological innovations in developing countries substantiate this statement (e.g., Qaim 2000). This increasing gap, as regards

Figure 3.2

The Green Revolution

In the early 1960s, the United Nations Food and Agriculture Organization (FAO), in collaboration with the CGIAR, launched the so-called Green Revolution. Central to the Green Revolution is the belief that investments in improving agricultural productivity are a prerequisite to initiating the process of economic development and that technology development is key to increasing agricultural outputs. An ambitious R&D program was established aiming at intensified production through the development of "high-yielding varieties" (HYVs), particularly of rice and wheat. These varieties gave improved yields, especially when sufficient amounts of water, fertilizer, and pesticides were added. This R&D policy was accompanied by a range of other policies, including provision of credit; supply and distribution of inputs (seed, water, fertilizer, pesticide); appropriate pricing policies for inputs and outputs; and infrastructure development.

The Green Revolution in rice and wheat was responsible for securing major yield increases in those grains. In the period 1960-90, global cereal production doubled, per capita food availability increased 37 percent, and real food prices declined 50 percent. In India, for example, wheat yields were raised by 50 percent and rice yields by 25 percent. This made India self-sufficient in wheat and rice.

This impressive global aggregate performance, however, masks considerable regional differences. The HYVs were mainly adopted in irrigated areas and favourable rain-fed areas. There was only limited adoption of HYVs in marginal regions that are prone to drought and/or flood – most areas in sub-Saharan Africa. The Green Revolution failed to produce cultivars that were relevant to the agro-ecological environments and socioeconomic conditions of many poor farmers. Of the small-scale, resource-poor farmers who did adopt the Green Revolution technologies – often attracted by packages of new seeds and inputs, along with credit to buy them – many were, sooner or later, faced with rising costs for inputs and declining prices for agricultural produce, which made it impossible for them to make sufficient profit. Many of them were forced to pay their debts by selling their land.

Three decades of Green Revolution agriculture have also led to other negative effects. The most often cited are pesticide and herbicide poisoning of people and animals, soil and water; soil erosion and land degradation; loss of genetic diversity; increased gap between the rich and the poor; and replacement of local economies and farming techniques with an export crop economy. Moreover, the dominance of scientists has shifted the role of farmers from original crop breeders to mere adopters of HYVs.

Sources: McCalla and Brown (2000); Serageldin and Persley (2000); Sexton, Hildyard, and Lohmann (1998).

yield and income, between poorer and richer farmers may lead to further marginalization of small-scale farmers and to an accelerated migration to the already overcrowded cities, where job opportunities are very limited. There may also be adverse effects on the social cohesion of these farming communities, and urban locations, due to these changing dynamics (see Mehta, this volume, Chapter 2).

The resulting "drop-out race" is further accelerated by another development enhanced by biotechnology – the increasing interchangeability of raw materials (Clark and Juma 1991; Leisinger 1999; Ruivenkamp 1989; Skerritt 2000). Recent developments in genetic modification and plant cell culture techniques allow the processing industry, which is mainly located in Western countries, to select the most attractive materials for its products. For example, high-fructose corn syrup (a sweetener produced from maize starch through genetically modified enzymes) has already captured over 10 percent of the world market of sweeteners, at the expense of sugar from sugar cane. As a result, millions of sugar cane farmers throughout the South have lost their source of income. Another example involves the potential replacement of coconut and palm oils by high-lauric-acid oil from GM canola (or spring rape), which is field-tested in Britain (Nottingham 1998; Skerritt 2000). The fragile economic positions and vulnerable farming systems of small-scale farmers imply that these farmers cannot anticipate such changing market demands rapidly enough, thereby leading to their further marginalization.

The grim picture sketched above does not imply that biotechnology has nothing to offer small-scale, resource-poor farmers. In contrast to Green Revolution technologies, biotechnology has the potential of tailor-making innovations for a wide range of end users, including currently ignored resource-poor farmers, in various environments. After all, biotechnology could reduce input use, reduce risk to biotic and abiotic stress, increase yields, and enhance (nutritional) quality – all traits that could enable the development of new crop varieties that are appropriate to poor producers and consumers (Broerse and Visser 1996; ECA 2002; Juma 2000; Persley 2000; Tripp 2001).

Moreover, transformation of small-scale agriculture in developing countries is often considered key in meeting the challenge of hunger and poverty (e.g., Broerse and Bunders 2000; ECA 2002; Krattiger 2000; McCalla and Brown 2000; Serageldin 2000). Small-scale agriculture is the farming system practised by more than 70 percent of farmers in the developing world (Reijntjes, Haverkort, and Waters-Bayer 1992). These farming systems are highly complex, with heterogeneous mixes of annual plants, livestock, and trees. The physical environment is characterized by fragile or problematic soils, water, wind, and temperature stresses. The commercial infrastructure and price ratios do not allow the widespread use of capital-intensive inputs. Furthermore, the lack of access to credit and technical assistance adds to the impediments. Consequently, productivity in small-scale agriculture is low. The area under small-scale agriculture is growing. Under structural adjustment programs, external inputs have become more expensive, with the result that fewer farmers can afford them. In order to survive, many small-scale, resource-poor farmers are forced to exploit the land beyond its carrying capacity. The sustainability of agriculture in these marginal lands poses a special challenge to researchers.

So how should we proceed? On the one hand, there is the broad applicability of biotechnology, but a research community hardly focuses research and development on poor producers and consumers. On the other hand, there is an urgent need in small-scale agriculture for new innovations. How can the supply (biotechnology innovations) appropriately meet the needs of the poor in developing countries? Tailoring biotechnological innovations for the needs of small-scale, resource-poor farmers is not easy or straightforward. The complex issues surrounding food provision are unlikely to be solved by a new technological fix. Poor people starve because they do not have access to information and to land on which to grow food, or do not have the money to buy food or inputs, or do not live in a country with a state welfare system (Sexton, Hildyard, and Lohmann 1998; Tripp 2001). Focusing on technology alone, while ignoring the underlying causes of hunger or inability to increase food supplies, may exacerbate rather than improve the situation.

If the potential value of biotechnology for poverty alleviation, food security, and sustainable development is to be realized, then there is a need for a double shift in the research paradigm (Bunders, Haverkort, and Hiemstra 1996; Serageldin and Persley 2000). First, efforts in biotechnology research and development should be specifically focused on agro-ecological systems and crops, livestock, fish, and trees important to poor people in developing countries. Second, research needs to be contextualized within the broader socioeconomic and cultural situations of the poor and within a deeper understanding of sustainability issues. In other words, the developed biotechnologies will have to be appropriate[4] for the more diverse and complex systems found, for example, in small-scale agriculture. The development of appropriate biotechnologies will require concerted efforts involving all relevant actors, including biotechnologists, farmers, consumers, nongovernmental organizations, government officials, and the private sector.

Interactive and Participatory Approaches

The notion to take socioeconomic, political, and cultural aspects into account and to involve the local community in technology development projects is nothing new; it goes back to the early 1970s (Cernea 1991;

Dusseldorp and Box 1990). As a result, people and institutions began experimenting with new approaches to the development of technological innovations.[5] Farming systems research (and extension) emerged as a multidisciplinary approach in which scientists tried to take whole production systems into account rather than focusing on selected aspects of them (Cornwall, Guijt, and Welbourn 1993). Applied agricultural research was expanded from on-station trials to include on-farm trials. In the multidisciplinary approach to the development of technological innovations, small-scale, resource-poor farmers were, for the first time, included in the R&D process, although more often as passive providers of land, labour, and information than as active participants.

However, a decade later, evaluation studies revealed that it is not sufficient to consult farmers only during the design phase of an R&D project. It was realized that farmers need to become active participants throughout the project, continuously providing checks and balances on information collected, priority setting, ideas on possible solutions, proposed sequence of actions, project evaluation, and so on (Cornwall, Guijt, and Welbourn 1993; Scoones and Thompson 1993). Based on this insight, many interactive and participatory approaches were developed, each with its own emphasis and field of application. To illustrate what these approaches entail, we describe one such approach – the interactive bottom-up (IBU) approach – in Figure 3.3. The IBU approach has been specifically designed to guide the development of biotechnological innovations for small-scale agriculture.

The idea that interactive and participatory approaches to biotechnological innovation processes are important is now increasingly acknowledged in both the developed and the developing world. Professor Swaminathan (1996, xi), director of the M.S. Swaminathan Foundation for Rural Development, has advocated the application of "a participatory and interactive methodology of research involving farmers, scientists and policy makers, as well as a broader process of institutional change." According to Swaminathan, this is the way to proceed in the field of biotechnology development for small-scale farmers. The Dutch Directorate General for International Cooperation (DGIC) is also convinced of the necessity of interactive and participatory approaches and is applying such approaches in its research and technology program (Storm 1997).

It is one thing, however, to acknowledge the power and validity of interactive and participatory approaches to the innovation process, yet another to apply such approaches sensitively and consistently. The growing awareness of the necessity of the application of these approaches is not easily operationalized in everyday practice in formal institutions. Evaluation studies on the implementation of participatory approaches in formal institutions show that various governments, agencies, NGOs, universities, and other organizations in the South have attempted to adopt participatory approaches

Figure 3.3

The interactive bottom-up approach

The IBU approach consists of a set of principles and guidelines. The principles lay the foundation for the approach and need to be strictly adhered to, while the guidelines provide phases, tools, and methods that can be applied flexibly depending on local circumstances. The IBU approach is based on the following seven principles:

1 The entire innovation process is centred on the vision that biotechnological innovations can contribute to small-scale agriculture. This vision provides a sense of direction, and central participants in the project are selected on the basis of their commitment to the vision.
2 Small-scale farmers play a prominent role in decision making throughout the innovation process. This is probably one of the best guarantees that resulting biotechnological innovations are tailored to the context, needs, and capabilities of the people who are supposed to benefit from them.
3 The development of trust relationships is facilitated. Trust is recognized as crucial if researchers applying the IBU approach are to obtain sensitive information and tacit knowledge.[1] It also encourages mutual learning and risk-taking behaviour. At the beginning of the process, the level of trust is generally low among actors with vested interests, since the aim is to induce changes.[2]
4 Mutual learning is facilitated. This is especially important for reaching consensus. It is important that each person involved in the process recognizes the others' expertise and potential contribution to the concerted effort and behaves accordingly.
5 Coalitions are built. A coalition of people from different areas serves as a platform for the necessary checks and balances through which the research process is corrected. Coalition building ensures that sufficient and appropriate support (endorsement, approval, legitimacy) and resources (knowledge, funds, materials, time) are made available to maintain momentum.
6 Different types of knowledge are integrated, particularly formal and informal knowledge and experiential knowledge of different stakeholder groups.
7 An interdisciplinary team manages the process. The team (1) has a mediating and facilitating role between the various actors, anticipating problems that result from differences in views, language, and power, (2) collects, exchanges, links, and integrates information and knowledge from many different sources and disciplines, and (3) acts as a change agent.

Roughly four phases can be distinguished within the IBU approach. The objectives of the *first phase* are to establish (and train) a team of researchers/practitioners and to become familiar with the socioeconomic, ecological, and political setting as well as the local community in a selected area. The *second phase* involves gathering information and knowledge on the issues concerned

►

◄ *Figure 3.3*

and the perspectives and views of relevant actors. In a more or less spiral-like way, actor perspectives are alternated, each time building on previous information and knowledge in an increasingly sophisticated way. At the end of this phase, the preliminary findings are laid down in an intermediary document. In the *third phase,* participants are brought together in interactive workshop settings to review and discuss intermediate findings, to achieve wide dissemination and legitimization, and to allow for the incorporation of new ideas and perspectives. These workshops are characterized by close interactions between participants, facilitating mutual feedback and the development of shared constructions. The plan of action, which resulted from the previous phase, forms the input to the *fourth phase,* in which specific projects are formulated and implemented. Project formulation and implementation can be undertaken at any level and may focus directly either on the local community or on research institutions or policy-making bodies, or any combination of these. As in the previous phases, it is essential that formulation and implementation should be interactive, exploring the options in close collaboration with all concerned. Each phase within the IBU approach consists of activities that are usually undertaken several times, leading to an iterative and dynamic process.

1 Tacit knowledge is highly personal and hard to formalize, making it difficult to communicate or to share with others. It is deeply rooted in an individual's actions, experiences, ideals, and values. It can be segmented into two dimensions: the technical dimension, which encompasses the kind of informal and hard-to-pin-down skills or crafts captured in the term "know-how," and a cognitive dimension that consists of schemata, mental models, beliefs, and perceptions so ingrained that we take them for granted (Nonaka and Takeuchi 1995, 8).

2 Accomplishing innovation and change invariably threatens the status quo and thus raises uncertainty, anxiety, and defensive behaviour in many actors (Pfeffer 1992). In addition, many actors in the context of biotechnology and small-scale agriculture are likely to have no history of interaction and are very diverse in terms of cultural background and interests. This also reduces the level of trust between various actors.

Sources: Broerse (1998); Broerse and Bunders (1999); Bunders, Haverkort, and Hiemstra (1996).

with varying degrees of success (Bainbridge 2000; Broerse and Bunders 2000; Chambers 1997; Gaventa 1998). Institutionalization of interactive and participatory approaches has turned out to be a complicated and difficult organizational learning process.

Analyses of the institutionalization of interactive and participatory approaches indicate that a lack of success is attributable to insufficiencies in participatory methodology, competency of practitioners, and institutional setting. First, the methodology used should include the basic principles of interactive and participatory approaches. Adherence to these principles puts specific demands on the methods applied throughout the innovation process. Second, it requires that the people involved in executing the approach (the practitioners) are capable and willing to apply these methods. And third, it is important that practitioners feel supported and rewarded within their broader institutional setting. Below we discuss each element on a conceptual level.

Participatory Methodology

For the successful development of appropriate biotechnology, research methods and tools should be in line with the principles of interactive and participatory approaches. The principles of the approach thus need to be made as explicit as possible. In addition, the possible methods and tools, which can be applied at the various stages in the innovation process, need to be well specified but should not be considered rigid blueprints. Just as technologies have to be tailored to their contexts of application, so too does an interactive and participatory approach to technology development have to be tailored to its context of application, including human resources and the institutional setting. Therefore, a participatory methodology should include, besides some general guidelines on how to structure activities, a range of methods and techniques for generating knowledge and interaction on which the practitioners can draw as required (Broerse and Bunders 2000; Chambers 1997, 2002; Zweekhorst 2003). The proclamation of applying a participatory approach to innovation has little meaning if the basic principles are in practice not seriously addressed; such a proclamation is then mere rhetoric.

Competency of Practitioners

The success of a participatory methodology depends greatly on how it is implemented. The competencies of the practitioners are crucial; it is the practitioners who need to adapt the methodology to its context of application and to decide on the next steps. The ideal practitioner is usually described as a person who is not only able to gather research from various academic and nonacademic sources but also able to build relationships with various social groups, listen, encourage, build trust, and facilitate dialogue, brainstorming, discussion, negotiation, mediation, and mutual learning. In addition, practitioners need to be able to assume multiple cognitive and social identities, to deal with ambiguity, to transcend disciplinary boundaries, and to reflect critically on their own actions and predispositions (Broerse and Bunders 2000; Chambers 1997, 2002; Fry 2003; Gibbons et al. 1994). Most people will not have been exposed to such a new research paradigm. Usually trained in a traditional monodisciplinary rote memorization system, the new practitioners face difficulties in understanding the participatory philosophy and lack the required competencies for effective application of the various techniques of participatory methodologies (Broerse and Bunders 2000; Chambers 1997; Tress, Tress, and Fry 2003). In the long run, training is key to bringing about the necessary changes in individual competency.

Institutional Setting

If interactive approaches are to be successfully institutionalized, it is important that the institutional structure and culture be appropriate – that is, they should support participation and collaboration both within the organization and with external contacts and clients. A supportive organizational setting should be flexible and decentralized, with working procedures and rules that allow the practitioners some freedom to experiment, to make (and learn from) mistakes, to respond to changing conditions and new opportunities, and to share information (Broerse and Bunders 2000; Chambers 1997; Gibbons et al. 1994). Such an atmosphere fosters inventiveness, creativity, and interaction. For many organizations, interactive and participatory approaches to technology development are still the exception rather than the rule. Particularly in more hierarchical organizations, this institutionalization process of interactive and participatory approaches is fraught with difficulties. It requires – besides staff training – changes in the way that projects are implemented and in the cultures and procedures of the organization (Blackburn and Holland 1998; Pretty and Chambers 1993).

Although institutions (or rather their people) can prove to be remarkably resistant to moves in the direction of more decentralized and heterarchical forms of organization and management (Howle, Neilson, and Ortiz 1996), change can be brought about (Pretty and Chambers 1993; Gaventa 1998; Zweekhorst, Broerse, and Bunders 2003). The incentive is usually the recognition, within the organization, that past approaches have failed and that there is a sense of urgency for change. This recognition and sense of urgency is combined with a commitment of top management and a growing number of staff to a new, more participatory, approach.

The above-mentioned elements are mutually reinforcing; the individual elements are starting points and preconditions, but none is likely to become established, be sustained, or spread well unless it receives support from all the other elements. This mutual reinforcement suggests a process in which methodological improvement, institutional change, and increased competency of practitioners succeed one another; every improvement in one element allows for and requires a new improvement in another element – no one element can be ignored. Consequently, a spiral process of learning and action develops, leading to increasing levels of sophistication in all elements (Zweekhorst 2003).

Conclusion

This chapter has dealt with current trends in biotechnology, particularly focusing on developing countries, and the question of how research and development can be realized from the perspective of the poor, such as small-scale, resource-poor farmers. Biotechnology is clearly not a magic bullet that can lead to poverty alleviation, food security, and sustainable development overnight. The reality of poverty is too complex for a simple technological fix; biotechnology research is but one factor among many that could have an impact on rural poverty, food production, and development. According to Spillane (2000, 1), "the potential contribution of biotechnology to developing country agriculture or to poverty alleviation is considered to have been overstated, in the short term at least."

Over the longer term, biotechnology may be a powerful tool in generating social, economic, and environmental benefits for developing countries. Provided that biotechnology development is specifically targeted at the poor, and an interactive and participatory approach is taken to guide the R&D process, this promise could be more fully exploited. The available evidence shows that putting these pro-poor policies and participatory research strategies in place is going to be a real challenge, requiring creative, forceful, visionary, careful, and sometimes daring actions. The alternative is to settle for the status quo. In that case, the introduction of ag-biotech innovations will probably lead to a widening gap between the rich and the poor and in that way is likely to threaten the stability and social cohesiveness of the developing world. We hope that the issues raised in this chapter will inspire scholars to make a difference and to help facilitate the generation of biotechnologies that may indeed be conducive to poverty alleviation, food security, and sustainable development and thus to increased social cohesiveness.

Notes

1 Intellectual property rights (IPR) is a broad term for the various rights granted by law for the protection of economic investment in creative efforts. A patent is an IPR granted by the government to inventors to exclude others from imitating, manufacturing, using, or selling a specific invention for commercial use during a certain period. This exclusivity arrangement contrasts with another IPR mechanism – plant breeders' rights. These rights are granted to plant breeders to exclude others from producing or commercializing material of a specific plant variety for a period of fifteen to twenty years. However, others may use the variety for research purposes (research and breeders' exemption), and farmers may use the harvested material of protected varieties for the next production cycle on their farms (farmers' privilege) (van Wijk, Cohen, and Komen 1993).

2 This is also a problem if the country does not have a national IPR system, because in that case research institutes tend to use Material Transfer Agreements (MTAs) – a legal agreement formalized between the partners involved that provides the supplier of biological material with sufficient protection while facilitating the freedom necessary for research. These agreements also require legal expertise in drawing them up or assessing them.

3 See <http://www.isaaa.org>.

4 A technology can be considered "appropriate" when it is adapted to the prevailing conditions in the area in which it is to be applied, making the best use of local resources and being economic – with regard to costs, risks, and values – to users in comparison with available alternatives.

5 The conventional approach to the development of technological innovations was a centralized, top-down approach – also called the "linear model" (Williams and Edge 1996). Universities and research institutes were considered the principal sources of new technology. In their laboratories – where conditions are homogeneous and completely under control – new or improved technologies are developed by highly trained technical experts.

Then the technologies spread via extension services to end users in the peripheries (Röling 1991; Scoones and Thompson 1993).

References

Alvarez-Morales, A. 2000. "Mexico: Ensuring Environmental Safety While Benefiting from Biotechnology." In G.J. Persley and M.M. Lantin, eds., *Agricultural Biotechnology and the Poor: An International Conference on Biotechnology*, 90-96. Washington, DC: CGIAR.

Bainbridge, V. 2000. *Transforming Bureaucracies: Institutionalizing Participation and People Centred Processes in Natural Resource Management – an Annotated Bibliography*. London: International Institute for Environment and Development.

Biber-Klemm, S. 2000. "Biotechnology and Traditional Knowledge: In Search of Equity." *International Journal of Biotechnology* 2 (1/2/3): 85-102.

Blackburn, J., and J. Holland. 1998. *Who Changes? Institutionalizing Participation in Development.* London: Intermediate Technology Publications.

Braunschweig, T. 2000. *Priority Setting in Agricultural Biotechnology Research: Supporting Public Decisions in Developing Countries with the Analytic Hierarchy Process*. ISNAR Research Report 16. The Hague: ISNAR.

Broerse, J.E.W. 1998. *Towards a New Development Strategy: How to Include Small-Scale Farmers in the Biotechnological Innovation Process*. Delft: Eburon Publishers.

Broerse, J.E.W., and J.F.G. Bunders. 1999. "Pitfalls in Implementation of Integral Design Approaches to Innovation: The Case of the Dutch Special Program on Biotechnology." In C. Leeuwis, ed., *Integral Design: Innovation in Agriculture and Resource Management*, 245-65. Mansholt Studies 15. Wageningen and Leiden: Mansholt Institute-Backhuys Publishers.

–. 2000. "Requirements for Biotechnology Development: The Necessity of an Interactive and Participatory Innovation Process." *International Journal of Biotechnology* 2 (4): 275-96.

Broerse, J.E.W., and B. Visser. 1996. "Assessing the Potential." In J. Bunders, B. Haverkort, and W. Hiemstra, eds., *Biotechnology: Building on Farmers' Knowledge,* 131-80. London: Macmillan Education.

Bunders, J.F.G., and J.E.W. Broerse, eds. 1991. *Appropriate Biotechnology in Small-Scale Agriculture: How to Reorient Research and Development.* Wallingford: CAB International.

Bunders, J.F.G., B. Haverkort, and W. Hiemstra, eds. 1996. *Biotechnology: Building on Farmers' Knowledge*. London: Macmillan Education.

Bustamante, P.I., and S. Bowra. 2002. "Biotechnology in Developing Countries: Harnessing the Potential of High-Tech SMEs in the Face of Global Competition." *Electronic Journal of Biotechnology* 5 (3) <http://www.ejbiotechnology.info/content/vol5/issue3/>.

Cernea, M.M. 1991. "Knowledge from Social Science for Development Policies and Projects." In M.M. Cernea, ed., *Putting People First: Sociological Variables in Rural Development,* 2nd ed., 1-41. New York: Oxford University Press.

Chambers, R. 1997. *Whose Reality Counts? Putting the First Last.* London: Intermediate Technology Publications.

–. 2002. *Participatory Workshops: A Sourcebook of 21 Sets of Ideas and Activities*. London: Earthscan Publications.

Clark, N., and C. Juma. 1991. *Biotechnology for Sustainable Development: Policy Options for Developing Countries*. Nairobi: ACTS Press.

Cohen, J.I., ed. 1999. *Managing Agricultural Biotechnology: Addressing Research Program Needs and Policy Implications*. Biotechnology in Agriculture Series 23. Wallingford: CAB International.

Cohen, J.I., C. Falconi, and J. Komen. 1998. *Strategic Decisions for Agricultural Biotechnology: Synthesis of Four Policy Seminars*. ISNAR Briefing Paper 38. The Hague: ISNAR.

Cohen, J.I., C. Falconi, J. Komen, and M. Blakeney. 1998. *Proprietary Biotechnology Inputs and International Agricultural Research*. ISNAR Briefing Paper 39. The Hague: ISNAR.

Cornwall, A., I. Guijt, and A. Welbourn. 1993. *Acknowledging Process: Challenges for Agricultural Research and Extension Methodology.* IDS Discussion Paper 333. Brighton: Institute Development Studies.

Dawe, D., R. Robertson, and L. Unnevehr. 2002. "Golden Rice: What Role Could It Play in Alleviation of Vitamin A Deficiency?" *Food Policy* 27: 541-60.

Dusseldorp, D.B.W.M., and L. Box. 1990. "Role of Sociologists and Cultural Anthropologists in the Development, Adaptation, and Transfer of New Agricultural Technologies." *Netherlands Journal of Agricultural Sciences* 38 (4): 697-709.

Dutfield, G. 2000. "Biodiversity in Industrial Research and Development: Implications for Developing Countries." *International Journal of Biotechnology* 2 (1/2/3): 103-14.

ECA. 2002. *Harnessing Technologies for Sustainable Development.* Economic Commission for Africa Policy Research Report. Addis Ababa: ECA.

Falconi, C.A. 1999. "Agricultural Biotechnology Research Indicators and Managerial Considerations in Four Developing Countries." In J.I. Cohen, ed., *Managing Agricultural Biotechnology: Addressing Research Program Needs and Policy Implications,* 24-37. Biotechnology in Agriculture Series 23. Wallingford: CAB International.

Fry, G. 2003. "On Needs for Training." In B. Tress, G. Tress, A. van der Valk, and G. Fry, eds., *Interdisciplinary and Transdisciplinary Landscape Studies: Potential and Limitations.* Delta Series 2. Wageningen, The Netherlands.

Gaventa, J. 1998. "The Scaling-Up and Institutionalization of PRA: Lessons and Challenges." In J. Blackburn and J. Holland, eds., *Who Changes? Institutionalizing Participation in Development.* London: Intermediate Technology Publications.

Gibbons, M., C. Limoges, H. Nowotny, S. Schwartzman, P. Scott, and M. Trow. 1994. *The New Production of Knowledge: The Dynamics of Science and Research in Contemporary Societies.* London: Sage Publications.

Howle, E., G.L. Neilson, and D.J. Ortiz. 1996. *Creating Temporary Organizations for Lasting Change.* Lewiston: Booz-Allen and Hamilton.

IFPRI. 1997. *The World Food Situation: Recent Developments, Emerging Issues, and Long-Term Prospects.* Washington, DC: International Food Policy Research Institute (IFPRI).

James, C. 2002. *Global Status of Commercialized Transgenic Crops: 2002.* ISAAA Briefs 27. Ithaca, NY: ISAAA.

James, C., and A. Krattiger. 1999. "The Role of the Private Sector. Brief 4." In G.J. Persley, ed., *Focus 2: Biotechnology for Developing-Country Agriculture: Problems and Opportunities.* Washington, DC: International Food Policy Research Institute (IFPRI).

Juma, C. 2000. "Biotechnology in the Global Economy." *International Journal of Biotechnology* 2 (1/2/3): 1-6.

Komen, J. 1999. "International Collaboration in Agricultural Biotechnology." In J.I. Cohen, ed., *Managing Agricultural Biotechnology: Addressing Research Program Needs and Policy Implications,* 110-27. Biotechnology in Agriculture Series 23. Wallingford: CAB International.

Komen, J., J. Mignouna, and H. Webber. 2000. *Biotechnology in African Agricultural Research: Opportunities for Donor Organizations.* ISNAR Briefing Paper 43. The Hague: ISNAR.

Krattiger, A. 2000. "Food Biotechnology: Promising Havoc or Hope for the Poor." *Proteus* 17: 38-45.

Leisinger, K.M. 1999. *Disentangling Risk Issues, Biotechnology for Developing-Country Agriculture: Problems and Opportunities, Brief 5.* Washington, DC: International Food Policy Research Institute (IFPRI).

McCalla, A.F., and L.R. Brown. 2000. "Feeding the Developing World in the Next Millennium: A Question of Science?" In G.J. Persley and M.M. Lantin, eds., *Agricultural Biotechnology and the Poor: An International Conference on Biotechnology,* 32-36. Washington, DC: CGIAR.

Macer, D.R.J., J. Azariah, and P. Srinives. 2000. "Attitudes to Biotechnology in Asia." *International Journal of Biotechnology* 2 (4): 313-32.

Madkour, M.A., A.S. El Nawawy, and P.L. Traynor. 2000. *Analysis of a National Biosafety System: Regulatory Policies and Procedures in Egypt.* ISNAR Country Report 62. The Hague: ISNAR.

Manicad, G., and S. McGuire. 2000. "Supporting Farmer-Led Plant Breeding." *Biotechnology and Development Monitor* 42: 2-7.

Njobe-Mbuli, B. 2000. "South Africa: Biotechnology for Innovation and Development." In G.J. Persley and M.M. Lantin, eds., *Agricultural Biotechnology and the Poor: An International Conference on Biotechnology,* 115-17. Washington, DC: CGIAR.

Nonaka, I., and H. Takeuchi. 1995. *The Knowledge Creating Company: How Japanese Companies Create Dynamics of Innovation.* New York: Oxford University Press.
Nottingham, S. 1998. *Eat Your Genes.* London: Zed Books.
Persley, G.J. 2000. "Agricultural Biotechnology and the Poor: Promethean Science." In G.J. Persley and M.M. Lantin, eds., *Agricultural Biotechnology and the Poor: An International Conference on Biotechnology,* 3-21. Washington, DC: CGIAR.
Pfeffer, J. 1992. *Managing with Power: Politics and Influence in Organizations.* Boston: Harvard Business School Press.
Pingali, P.L., and G. Traxler. 2002. "Changing Locus of Agricultural Research: Will the Poor Benefit from Biotechnology and Privatization Trends?" *Food Policy* 27: 223-38.
Pinstrup-Anderson, P., and R. Pandya-Lorch. 2000. "Meeting Food Needs in the 21st Century: How Many and Who Will Be at Risk?" Paper presented at AAAS anuual meeting, February, Washington, DC.
Pretty, J.N., and R. Chambers. 1993. *Towards a Learning Paradigm: New Professionalism and Institutions for Agriculture.* Discussion Paper 333. Brighton: Institute of Development Studies.
Qaim, M. 2000. "Biotechnology for Small-Scale Farmers: A Kenyan Case Study." *International Journal of Biotechnology* 2 (1/2/3): 174-88.
Reijntjes, C., B. Haverkort, and A. Waters-Bayer. 1992. *Farming for the Future: An Introduction to Low-External-Input and Sustainable Agriculture.* London: Macmillan Education.
Röling, N. 1991. "Institutional Knowledge Systems and Farmers' Knowledge: Lessons for Technology Development." In G. Dupré, ed., *Savoir paysan et développement* [Farming Knowledge and Development], 489-514. Paris: Karthala-ORSTOM.
Ruivenkamp, G. 1989. *The Introduction of Biotechnology in the Agro-Industrial Production Chain.* Utrecht: Jan van Arkel Publishers.
Rybicki, E. 1999. "Agricultural Molecular Biotechnology in South Africa: New Developments from an Old Industry." *AgBiotechNet* 1 ABN 023.
Salazar, S., C. Falconi, J. Komen, and J.I. Cohen. 2000. *The Use of Proprietary Biotechnology Research Inputs at Selected Latin American NAROs.* ISNAR Briefing Paper 44. The Hague: ISNAR.
Sampaio, M.J.A. 2000. "Brazil: Biotechnology and Agriculture to Meet the Challenges of Increased Food Production." In G.J. Persley and M.M. Lantin, eds., *Agricultural Biotechnology and the Poor: An International Conference on Biotechnology,* 74-78. Washington, DC: CGIAR.
Scoones I., and J. Thompson. 1993. *Challenging the Populist Perspective: Rural People's Knowledge, Agricultural Research, and Extension Practice.* IDS Discussion Paper 332. Brighton: Institute of Development Studies.
Serageldin, I. 2000. "The Challenge of Poverty in the 21st Century: The Role of Science." In G.J. Persley and M.M. Lantin, eds., *Agricultural Biotechnology and the Poor: An International Conference on Biotechnology,* 25-31. Washington, DC: CGIAR.
Serageldin, I., and G.J. Persley. 2000. *Promethean Science: Agricultural Biotechnology, the Environment, and the Poor.* Washington, DC: CGIAR.
Sexton, S., N. Hildyard, and L. Lohmann. 1998. *Food? Health? Hope? Genetic Engineering and World Hunger.* Corner House Briefing 10. Dorset: Corner House.
Skerritt, J.H. 2000. "Genetically Modified Plants: Developing Countries and the Public Acceptance Debate." *AgBiotechNet* 2 ABN 040.
Spillane, C. 2000. "Could Agricultural Biotechnology Contribute to Poverty Alleviation?" *AgBiotechNet* 2 ABN 042.
Storm, G. 1997. "Introduction to Discussion." In *Selected Regional Issues for Agriculture and Agricultural Research: Papers Presented at the ISNAR Symposium,* 35-37. The Hague: ISNAR.
Swaminathan, M.S. 1996. "Preface." In J. Bunders, B. Haverkort, and W. Hiemstra, eds., *Biotechnology: Building on Farmers' Knowledge,* x-xii. London: Macmillan Education.
Swanson, T., and T. Goschl. 2000. "Genetic Use Restriction Technologies (GURTs): Impacts on Developing Countries." *International Journal of Biotechnology* 2 (1/2/3): 56-84.
Tanticharoen, M. 2000. "Thailand: Biotechnology for Farm Products and Agro-Industries." In G.J. Persley and M.M. Lantin, eds., *Agricultural Biotechnology and the Poor: An International Conference on Biotechnology,* 64-73. Washington, DC: CGIAR.

Tress, B., G. Tress, and G. Fry. 2003. "Potential and Limitations of Interdisciplinary and Transdisciplinary Landscape Studies." In B. Tress, G. Tress, A. van der Valk, and G. Fry, eds., *Interdisciplinary and Transdisciplinary Landscape Studies: Potential and Limitations*. Delta Series 2. Wageningen, The Netherlands.

Tripp, R. 2001. "Can Biotechnology Reach the Poor? The Adequacy of Information and Seed Delivery." *Food Policy* 26: 249-64.

van de Sande, T., G. Ruivenkamp, and S. Malo. 1996. "The Socio-Political Context." In J. Bunders, B. Haverkort, and W. Hiemstra, eds., *Biotechnology: Building on Farmers' Knowledge,* 181-98. London: Macmillan Education.

van Wijk, J. 2000. "Biotechnology and Hunger: Challenges for the Biotech Industry." *Biotechnology and Development Monitor* 41: 2-7.

van Wijk, J., J.I. Cohen, and J. Komen. 1993. *Intellectual Property Rights for Agricultural Biotechnology: Options and Implications for Developing Countries*. ISNAR Research Report 3. The Hague: ISNAR.

Williams, R., and D. Edge. 1996. "The Social Shaping of Technology." *Research Policy* 25: 865-99.

World Bank. 1997. *World Development Report 1997*. New York: Oxford University Press.

Zhang, Q. 2000. "China: Agricultural Biotechnology Opportunities to Meet the Challenges of Food Production." In G.J. Persley and M.M. Lantin, eds., *Agricultural Biotechnology and the Poor: An International Conference on Biotechnology,* 45-50. Washington, DC: CGIAR.

Zweekhorst, M.B.M. 2003. *Institutionalizing an Interactive Approach to Technological Innovation: The Case of the Grameen Krishi Foundation*. Amsterdam: Printpartners IpsKamp.

Zweekhorst, M.B.M., J.E.W. Broerse, and J.F.G. Bunders. 2003. "Capacity Building for Community Development: The Experience of a Bangladeshi NGO." In *Strengthening Community Competence: Asia Pacific Perspectives*. Phitsanulok: Naresuan University.

4
Legitimation Crisis: Food Safety and Genetically Modified Organisms

Christopher K. Vanderpool, Toby A. Ten Eyck, and Craig K. Harris

Complacency and Contentment

For the past century in the United States, concerns with the food supply have challenged the social fabric of society, leading to the development of governmental bodies to protect the legitimacy of the food system. For instance, the Food and Drug Administration (FDA) and the Food Safety Inspection Service in the United States Department of Agriculture (USDA) were created in response to concerns about food safety and adulteration in the late nineteenth and early twentieth centuries. Once these agencies were created and shown to be useful in some respects, the general public in the United States assumed that they were doing their "jobs" – that is, keeping the food we eat safe and keeping contaminants out of the food supply chain. For example, the mandated widespread implementation of the pasteurization of milk greatly diminished food-borne illness from milk (Ten Eyck and Williment 2004). Improved sanitation in the production of poultry, pork, and beef, increased strictness of inspection in meat-processing plants, and expanded use of antibiotic preservatives in processed meat – all diminished the threats of zoonotic diseases (Schmidt and Rodrick 2002). The increased reliance on commercially canned and frozen fruits and vegetables reduces the risks of food-borne illness from home canning or unsanitary distribution systems.

In the 1990s, the US public was jolted from its complacency by a series of incidents of food contamination, ranging from agrichemicals to novel pathogens (Fox 1997). In 1990, it was widely claimed that the use of alar on apples posed a significant danger to the health of infants and children. In 1992-93, four children died from eating hamburgers that had been purchased at a fast-food chain and that were contaminated with a formerly obscure pathogenic form of the common microbe E. coli. In 1996, two children died from drinking commercially produced fruit juice that had not been pasteurized and that contained the same pathogen. At the same time, it was becoming increasingly apparent that the UK beef production system was contaminated by a previously unknown form of a transmissible spongiform encephalopathy – "mad cow disease." The final years of the twentieth century in the United States featured recurrent outbreaks of E. coli in sprouts, listeria in processed meats, salmonella in eggs, cryptosporidia in drinking water, and cyclospora in raspberries, and the US agrifood system was increasingly perceived as hazardous at this time. Enter a novel type of product – genetically modified (GM) foods – and the stage is set for a continued breach of confidence between government and industry and the population at large. As we will discuss in more detail below, there is not a generally accepted boundary between foods produced by traditional methods and foods produced by biotechnologies. However, the introduction of the Flavr Savr™ tomato in 1995 may be considered the beginning of the GM revolution. From a tomato that had been engineered for greater longevity, the revolution moved on to beans and corn that were not harmed by the direct application of herbicide, corn and cotton that incorporated a gene from a species of bacteria to produce an insecticidal substance, salmon that included a gene to increase their rate of growth by 200 percent, and rice with elevated levels of beta carotene (a precursor of vitamin A) and iron.

Many segments of the public were highly skeptical of the ability of the US regulatory system to safeguard the public interest in considering the introduction of these novel agrifood products. If the regulatory system could not protect US consumers from food-borne illnesses caused by agrichemicals that had been approved by the system itself, and by pathogens that had been around for many decades and that the regulatory system had been designed to control, how could this system be expected to make effective judgments about GM products for which there were no analogues in previous experience?

It is this tension between the historical faith in the efficacy of the US regulatory system and the deep doubt about the wisdom of judgments concerning the agrifood biotechnologies that is the focus of our chapter. This tension implicates not only government but also industry and science in their claims to act on behalf of the public good. Mass and professional media and nongovernmental organizations are also implicated in their role of heightening public awareness of the issues surrounding agrifood biotechnologies.

In this chapter, we develop a theoretical model of the linkages between science, industry, and government in agrifood systems and the disconnections that have occurred between the agrifood system and the cultural frameworks in which food and society issues are embedded. These disconnections have resulted in a crisis of legitimacy that is characterized by a colonization of everyday life by government and regulatory systems that are not living up to expectations. We conclude by discussing some of the ways in which citizens and consumers can act to reclaim some control.

Legitimation Crisis: Risk and the Food System

Legitimation and Its Crises

The linkage of GMOs and food safety points to a crisis of legitimation, and a crisis of rationality, which have several interlocking dimensions. Our understanding of legitimation follows closely the works of Max Weber (1968) and Jürgen Habermas (1973). For Weber, legitimation means that a government is viewed as having the "right" to expect compliance and to demand allegiance to its authoritative decisions. In Habermas's view, the effective functioning of government requires both the ability to plan effectively and the perception, by stakeholders, that government activity is legitimate; the perception of legitimacy leads to the mass loyalty and compliance necessary for the effective functioning of society (McCarthy 1984). For Habermas, a failure of the administrative planning function is a crisis of rationality in which the ability of the administrative system to make appropriate rational decisions becomes problematic (49). In a legitimation crisis, when a government is no longer perceived as having authority, people are not motivated to support the claims made by the state. In practice, it is difficult to distinguish between these two forms of crisis. As Thomas McCarthy notes, "a disorganization of steering performances ... leads to a withdrawal of legitimation" (367). If one perceives the national government as having failed to deal effectively with food-borne illness, then one may withdraw one's loyalty and compliance to that aspect of regulations and governance, though other spheres of governmental control may be left intact. At the same time, if one perceives the national government as completely illegitimate, then one may also perceive the government as incapable of effective administration. These two crises become intertwined in the discourse about GMOs as opponents attempt to portray the government as incompetent and as proponents attempt to portray the government as benign and opponents as malcontents.

For Habermas (1973, 46-47), a legitimation crisis occurs when a legitimizing system does not succeed in maintaining loyalty. Modern economic and political systems experience crises on nearly a regular or cyclical basis characterized respectively by an economic failure in generating a sufficient and sustained level of wealth and a rationality crisis marked by a failure to deal effectively with the problems of the economic system. Such crises result in a disorganization of the state. At such times, there is an increasing gap between the expectations of citizens and the ability of the economic and political system to fulfill them (73). A feeling increasingly emerges that the economy and the polity are inauthentic. The result is an identity crisis in which cultural traditions are undermined and weakened. As the identity crisis accelerates, there is an increasing alienation of the people. The media that serve as a general mechanism of production and reproduction of meanings and values in society (i.e., culture) become a vehicle for exposing distortions in the economy and polity. The legitimacy of the state and its regulatory bodies, and of institutions linked to the state, to maintain and establish administrative and normative structures is undermined. Under such conditions, the public may develop alternative meanings and values that challenge the worldviews of dominant institutions.

Legitimation crises represent a challenge to economic and political systems. The response of these systems to such crises is to intensify the rationalization of society. For Habermas (1987), the rationalization processes dominant in the world of economic and political systems and their administrative structures begin to penetrate the sociocultural world of everyday life. The latter is the "lifeworld" that is the realm of social relationships and sociocultural values that are the foundation of identity. In the sociocultural system reside the communicative processes, structures, and relationships that generate meanings and values that enable individuals and groups to develop a sense of a take-it-for-granted reality. If this does not happen, trust cannot develop, and society disintegrates. The "systemworld," in contrast, is marked by commodity production and instrumental rationalization as the basis of social organization. The systemworld seeks to overcome its delegitimation by an increasing penetration of the lifeworld through rationalization processes – the values of the systemworld become the values of the lifeworld.

The public begins to voice opposition to the structures and processes that dominate social life and may even try to "exit" by seeking an alternative set of pathways and institutional arrangements to meet its needs (Hirschman 1970). In modern societies, legitimation crises are endemic because of the private appropriation of wealth and its unequal distribution, the suppression of human interests, and the failure adequately to integrate individuals and groups in society through democratic discourse (McCarthy 1984, 358). Those endemic crises are especially poignant when they are focused on human food issues, to which we now turn.

Food and Social Subsystems

For several reasons, food is an important dimension of legitimation in society. Not only is food required for subsistence and survival, but it also expresses the identities of cultural and ethnic groups (Caplan 1997). In many societies, the production of food engages the vast majority of society, both in the field and at home. For households that do not raise their own food, dependable access to affordable and acceptable foodstuffs is an important basis for the evaluation of the legitimacy of government and the social order.

In the GMO legitimation crisis, several systems are interlocked:

- *agrifood system,* from production and processing to distribution, including multinational corporations and agrifood industry;
- *cultural system* of values, knowledge, and trust, including the media system as a communicator of values and risks;
- *food safety system* of policies, regulations, certification, auditing, monitoring, and surveillance; and
- *scientific system* of university- and industry-based research and development as well as governmental scientists and technicians.

These systems echo the systems that Habermas (1973) saw as experiencing deepening systemic legitimation crises.

The agrifood system includes three basic forms of producers and processors: organic, industrial, and small scale (Albrecht 1997; Belasco 1993). All are driven by market forces and by their perceptions and understandings of what comprises an agrifood system. The dominant actors in GMOs are corporate and industrialized agricultural organizations that view agrifood production as structured by commodity and market formations (Bijman 1999; Hendrickson and Heffernan 2002). Organic and small-scale farmers are also market oriented but tend to place a greater importance on farming and/or organic food production as a way of life (see Mehta, this volume, Chapter 2). Yet the viability of all producers and processors depends on market forces (Belasco 1993), which in turn are shaped by cultural systems (DiMaggio 1990, 1994).

Cultural systems, as we are defining them, develop around sets of knowledge (e.g., religious institutions, mass-media outlets). This knowledge, in turn, is disseminated to and interpreted among a general public, who uses that knowledge to understand and navigate through social environments (Swidler 1986; Znaniecki 1952). The knowledge controlled within a cultural system is not equally distributed or normatively relative. Some knowledge patterns become more important than others and take on added significance to become values, meanings, and rules, which in turn can become the source of power within a cultural sphere (Wuthnow 1987). These patterns of knowledge can even move into the realm of the taken-for-granted, becoming the plausibility structures that inform us on whom and what we are to trust in given situations (Berger 1969). In addition to containing knowledge and plausibility structures within itself, each system involves communication processes that help to define, and are defined by, the system. These include experiential knowledge, public wisdom, and media discourse (Gamson 1992). As with knowledge, communication processes are not normatively relative. Some communication sources or channels are more trusted than others, and trusted processes can become tools of power.

Our argument here parallels that made by Peter Berger and Thomas Luckmann (1971). When people "know" an object, such as a familiar food, they trust that their knowledge, values, and beliefs can provide them with a satisfactory explanation and understanding of that reality. However, to the extent that a reality is less known, or if a reality that is unknown enters social life, there is an increasing sense of unease and distrust; we argue that this was the case with foods produced by previously unknown techniques of genetic engineering. As this disconcertedness becomes manifest, the social constructions that have been accepted as explaining reality are challenged. These include constructions that legitimated the belief that science, the state, and the economic system would do no "harm" to the public. In fact, it was expected that they would protect the public from external risks.

This tendency to distrust things that are new factors into the multivalence of the mass media. On the one hand, the same source carries conflicting information; stories about biotechnology usually contain a mixture of positive and negative attributions (TenEyck, Thompson, and Priest 2001). On the other hand, audiences generate varying interpretations of the same information; for example, despite the fact that the public has received similar information about the nature of biotechnology, different segments of the public interpret biotechnology as positive, negative, or neutral (Priest 2000; Priest and Gillespie 2000). This variance highlights the idea that we have become accustomed to relying on the media for information, but, because so much of the information does not resonate with our own experiential knowledge, we are not always inclined to trust it (Thompson 1995). In addition, levels of trust in information carried by the mass media seem to be issue specific (Ten Eyck 2000); for example, people seem to trust media coverage of medical biotechnology, while they question media coverage of food biotechnology (Durant, Bauer, and Gaskell 1998).

As we turn to the discourse about genetically modified foods, we assume that a majority of the lay public has very little experiential knowledge concerning genetic research (Grocery Manufacturers of America 2000). In addition, much of the technology being used is new, so public wisdom on the subject is underdeveloped (Ipsos-Reid Agri-Food Research Group 2000a). Instead, various publics must rely on information that is disseminated through their particular media channels. Lay persons rely on mass media, while professional persons rely on their more targeted media (Priest and Gillespie 2000). Interest groups may also seek guidance. Charles Seife (2001), for example, contends that the pope has been advised by the Pontifical Academy of Sciences to make an announcement that would significantly influence the debate over GM foods. This is not to say that audience members in such a situation are cultural dupes; it is simply to say that most people have standard operating procedures for gathering information. As will be made clear, the lack of knowledge concerning a technology that is entering a cultural system can lead to questioning both its appropriateness and the motivations of those backing the technology.

The third system engaged in the GMO legitimation crisis is the food safety system. In the United States, this system has developed gradually since the concerns about meat processing, milk adulteration, and fruit and vegetable contamination in the early 1900s. At the national level, responsibility for food safety is partitioned among the Food and Drug Administration and the Centers for Disease Control and Prevention in the Department of Health and Human Services, and the Food Safety Inspection Service and the Agricultural Research Service in the Department of Agriculture (Committee to Ensure Safe Food from Production to Consumption 1998). Together these agencies are responsible for establishing public policy about food safety, implementing regulations to carry out those policies, monitoring compliance with regulations and assessing the safety of the food supply, researching the factors that affect food safety, and educating actors in the food system about food safety. In carrying out these functions, the agencies rely on many partners in state and local governments, in land grant institutions, and in various sectors of industry.

As noted above, the food safety system itself has been the object of significant questioning. Concerns about the dispersion of responsibility and lack of coordination led the US Congress to mandate a study by the National Academy of Sciences (Committee to Ensure Safe Food from Production to Consumption 1998), which concluded that the functioning of the system could be improved. It has been suggested that the food safety system illustrates well the tendency for agencies to be captured by the groups that they are supposed to be administering. One of the enduring criticisms of the system is that it has very little scientific apparatus of its own and thus has to rely on the science produced by the proponents seeking regulatory approval of new technologies (Turner 1970).

While the cultural system is key in the dissemination and interpretation of information, the science system is important in developing information as well as offering instructions on how that information should be used (Salter 1988). According to Robert Wuthnow (1987, 288), this role became legitimated and practically hegemonic as it developed in conjunction with the Protestant ethic and the capitalistic economic system of modern society. The state has also had a role to play as scientists have come to dominate many governmental offices. Those scientists who have not been directly employed by the state often seek public funding, developing linkages to the state through granting agencies. Habermas (1973, 84) does state that science can be (and has been) "criticized and convicted of residual dogmatism" by the state and other institutions, though he recognizes that governmental technocrats are needed to both legitimize governmental affairs and offer remedies for crises as they appear. Until a plausible alternative is developed, these are experts who are sought to protect the public against threats.

Food and Society

GMOs fit into two cultural spheres in the United States: technology and food. This is not the first time that food and technology have been combined. At the turn of the twentieth century, a controversy arose over the pasteurization of milk, with opponents arguing that pasteurization was an expensive process used to mask unsanitary dairy conditions (Larsen and White 1913). Currently, the same arguments can be heard against food irradiation (Ten Eyck 1999). In fact, many pre- and postharvest technologies have been questioned, such as pesticides, animal drugs, and food additives.

Genetic engineering in crops, much like hormones in animal production, pesticides in fruit and vegetable production, irradiation in storage, and pasteurization in postharvest processing, incorporates the technology into the food, which makes it difficult to remove any trace or undo the effects of the technique. This characteristic of GMOs makes them more culturally sensitive in terms of the meanings, values, and trust in the food arena as compared to other technologies that can be more easily eliminated or undone. Food is an important aspect of cultural boundaries and self-identity (Bell and Valentine 1997; Counihan and Van Esterik 1997; Shortbridge and Shortbridge 1998). The introduction of a new technology into a deeper level of this cultural sphere may be seen as a threat to these boundaries and identity markers (Habermas 1989, 188-265). This forces claims makers to create affirmations of legitimacy for one side or the other – that is, one side's safe, bioengineered food and food products become the other side's "Frankenfood."

One of the reasons that discourse about GMOs has seemed so rancorous (Brown 2000), and concern about GMOs has exploded into a crisis of legitimation, is that the proponents have not addressed the cultural resonance of food (Hesman 2001). In the case of GMOs, biotechnology corporations and agrifood processors and producers have put forward claims that do not correspond to food habits, meanings, and preferences. Typical claims involve environmental benefits (reduced pesticide usage, less soil erosion), economic benefits (higher yields, lower costs), or personal benefits for farmers (less labour time, less risk of pest damage) but nothing in terms of producing a healthier and higher-quality food product that fits consumer definitions of common food practices and food quality. Even the seemingly counter examples (increased iron and beta carotene in rice) are not culturally resonant in the societies for which they are intended (Pollan 2001).

The agrifood system, with its emphasis on market forces, failed to consider that in the cultural system the important issues are

- Is it food?
- Is it good for us?
- Is it safe?
- Can we trust it?

Many advocacy groups provide answers to these questions for GMOs: they are not natural and are not food but an artificial creation of the laboratory (De Visser et al. 2000; Teitel and Wilson 1999). They are not healthy; for example, introducing a gene may create a food allergen posing hidden dangers to consumers (Ipsos-Reid Agri-Food Research Group 2000b). GMOs are not safe since they may be producing new toxic substances (Brown 2000). Corporations and regulatory bodies cannot be trusted because financial gains rather than safe and healthy food are the motivation behind the fast introduction and spread of GMOs (Corporate Watch 2000; Vaidyanathan 2000). They not only threaten all food consumers but also pose specific threats to the sustainability of small-scale and organic farming (Bailey 1998; Dawkins 1997). Labels such as Frankenfood capture this cultural schism and are mentioned prominently in advertisements appearing in newspapers supported by groups such as the Sierra Club, Mothers and Others for a Livable Planet, Mothers for Natural Law, Organic Consumers Association, Center for Food Safety, and Greenpeace. The power and cultural imagery of these groups pose formidable opposition to biotechnology and its widespread use in society.

Legitimation Crisis

The schism between the cultural system and bioengineering and genetic modification has generated a lack of trust in the agrifood, food safety, and scientific systems. Consumers did not have detailed knowledge and understanding of biotechnology (Ipsos-Reid Agri-Food Research Group 2000c). The keepers of knowledge in industry and science provided only selective understandings and knowledge of what was being bioengineered and how the biotechnology process works (Brown 2000; Levidow and Carr 2000). Furthermore, the food safety system appeared to be in collusion with the agrifood and scientific systems in developing regulations that supported the introduction of bioengineered food and food products rather than developing policies and tests that would lead to adequate safety certification of GMOs (Levidow 2000; Littlejohns 2000; Vogt and Parish 1999). The resistance to labelling of GMO food and food products by multinational biotechnology companies, agricultural producers, and food processors with the agreement of the FDA and the USDA adds fuel to the controversy (Goldman 2000). Anti-GMO forces insist that all food and food products be labelled as containing GMOs or as GMO free (International Food Information Council 2001). They argue that such labelling is essential to protect the safety of the public from the hazards of GMOs, such as allergens and unforeseen GMO-related food safety problems (Kamaldeen and Powell 2000).

Following Habermas's theory of communicative action (1987), we argue that the systemworld of food producers and processors, of the state and its regulatory agencies, of industry and the economy, and of science and technology has colonized the lifeworld of consumers and their everyday, taken-for-granted reality of food habits, practices, beliefs, and trust. And this colonization occurred without any announcement. The consumer was passively to accept the presence of GMOs in the food chain and the assurance of safety provided by industry, science, and the state (GeneWatch UK 2000). Moreover, the food safety system is a contributor to this colonization by developing the social construction of safe food via its regulatory policies (Vogt and Parish 1999) and the creation of grades and standards (Busch 1997). This attempt to colonize the lifeworld has led to an important gulf between scientists and regulators on the one hand and consumers on the other (Gaull 1998). Scientists and regulators define healthy and safe differently from the cultural definitions of health and safety shared widely by consumers (Priest and Gillespie 2000).

Public reactions are forcing agribusiness (Monsanto UK 2000) and regulatory agencies to discuss GMO issues publicly. This is particularly problematic for regulatory agencies that normally take precautionary stances (Gunter and Harris 1998). All sides in the GMO controversy are following the standard formula in US society that is inclined to view controversial issues from current media logic: that is, issues are constructed around quick, black-and-white sound bites (Altheide and Snow 1991).

Moreover, the agrifood system engages in actions that are designed to delegitimate consumer concern. Arguments are made that consumers and critics lack knowledge, have personal characteristics such as paranoia (No More Scares 2000), or possess wrong-minded cultural values. Consumers and critics are "new Luddites" who want to break the movement of scientific progress in order to return to a prescientific past characterized by minimalist technology. The critics are declared to be irrational and anti-scientific. This is an attempt to delegitimate consumer concern by saying that consumers lack knowledge or have "wrong-minded" cultural values. In counterpoint, consumers and critics delegitimate the food safety system for not doing enough to protect the public from food-borne illnesses and hazards. They increasingly believe that the food safety system lacks authenticity as they begin to believe that it worked in cooperation with the agrifood system to prevent an open discussion and debate on the safety of the food products deriving from agribiotechnology. The food safety system promises to protect the public from harm yet works with agrifood industry to thwart efforts to develop further safety mandates and regulations to ensure that GMOs are safe to eat (Wright 2001). Consumers delegitimate the food safety system by arguing that not enough is being done to guarantee the safety of GMOs. The technoscience of agribiotechnology appears to be out of control. Food safety regulations appear to be corrupted and ineffective. The response of the lifeworld is to maintain its own well-being despite the opposition of the systemworld.

Institutions and the Social Order

As with many technologies (Clarke and Montini 1993; Nelkin 1996), knowledge of genetically modified foods has been constructed and developed within organizations and across institutions. While the types of organizations and institutions involved in this process are numerous, we would like to focus briefly on four such entities: universities, industry, governmental agencies, and popular mass media. Although this categorization scheme is used only as a heuristic device, it does offer the opportunity to explore the control efforts of different actors within the system.

Scholarly literature on the role of universities as research centres is quite extensive. Andrew Pickering (1995), for example, described and analyzed the development of the bubble chamber (used in experimental elementary particle physics) at the University of Michigan and the University of California at Berkeley, while Harry Collins and Trevor Pinch (1993) did the same for work on cold fusion at the University of Utah and the sex lives of whiptail lizards at the University of Texas. Andrew Isserman (2000) has done the same for the development of biotechnology research at the University of Illinois. In these cases, as well as various other studies, the conclusions typically drawn are that universities serve as centres where unique and controversial research can be given an opportunity to reach either a positive or a negative conclusion and that the knowledge developed through this research is a process of negotiations. Scientific facts are not discovered but constructed through interactions between scientists, research assistants, animals, plants, and nonliving elements. In addition, not all results are felt to be worthy of publication (Latour and Woolgar 1979), and some reported results are nothing more than reflections of researchers' needs (Bridgstock 1982; Hogan 1997).

In the area of biotechnology and universities, two main lines of inquiry have arisen. First, the potential for significant profits has led to concerns over the objectivity of university researchers in this area, especially when research is funded by large agribusinesses (Etzkowitz 1996). Second, there has been concern over intellectual property rights as biotechnology companies and university scientists must come to some agreement on the division of profits from research (Thompson 1997). These concerns over boundaries and profits may impact the exchange of knowledge between actors from these types of organizations, in turn influencing the knowledge that becomes publicly available (Kloppenburg 1991). Whatever the outcome of these potentially confrontational situations, studies have found more linkages being developed between university scientists and private industry (Krimsky, Ennis, and Weissman 1991; Liebeskind et al. 1996). At the same time, these linkages have become more contested; an arrangement between Novartis and the University of California, Berkeley, whereby the corporation would provide the Department of Plant and Microbial Biology with $25 million over five years, aroused a great deal of controversy and has not been renewed.

These associations between universities and industry do not indicate that industry has abstained from producing its own knowledge in the area of genetically modified foods. In fact, companies such as Monsanto and Pioneer Hi-Bred have been actively pursuing new areas of research that have shown potential marketability (Massieu and Yolanda 1991). In addition to developing specific plants and other agricultural products, these companies have been active in developing communication strategies to influence public opinion and stifle opposition (Kleinman and Kloppenburg 1991).

Marketing new food and agricultural products in the United States involves approval processes that are regulated by governmental agencies such as the United States Department of Agriculture and the Food and Drug Administration (Uchtmann and Nelson 2000). While these agencies do, at times, create novel or unique knowledge concerning food and drug items, our focus is on their regulatory function. This responsibility consists of making a ruling concerning the safety of a new food or technology as well as deciding not to take a stand on some issues. Whatever path is taken, the decision often has as much impact on consumer acceptance as does the scientific knowledge that made the product possible (Powell and Leiss 1997).

Each of these organizations creates knowledge, though the end result is typically a product of interorganizational negotiation. The final organizational actors of interest are the ones involved in disseminating the information coming from these other organizations. According to Douglas Powell and William Leiss (1997), the organizations that play a leading role in the dissemination of knowledge about these issues are the popular mass media. We do not advocate an approach that places the entire burden on the mass media, as studies have shown that members of the lay public construct their understandings of issues through various sources (Gamson 1992; Hajer 1995). Instead, we view the popular mass media as an important vehicle of information that also plays a part in setting the social agenda for politics and the public (Dearing and Rogers 1996).

At this point, it is important to make a distinction between knowledge and understanding. Knowledge, according to Lawrence Hazelrigg (1989, 122), "requires justification and warrant." This definition leads us to assert that knowledge is most often a narrow statement about a specific relationship between an object that is known and a thinking subject that wants to know it. This does not entail relationships between object, subject, and other objects and subjects. Instead, contextual statements are understandings (Chomsky and Raskin 1987).

This distinction between knowledge and understanding is more than an exercise in semantics. To analyze the roles of the organizations discussed above, we must understand what they are producing. While each organization generates knowledge and understanding, we argue that the level of understanding typically increases from the university to industry to regulatory agencies to the mass media. Knowledge production, on the other hand, decreases. So, while university researchers are interested in very specific characteristics of plants and gene manipulations, industry moves beyond this knowledge to think about how the product will play out in the market. Regulatory agencies, on the other hand, are not as concerned about marketability as about safety and acceptability. Reporters for the mass media then take these factors and try to make sense of them in a larger context (develop an understanding of how these new products and decisions will impact society) while generating little or no knowledge concerning the specific techniques of biotechnology. These various levels of knowledge and understanding can lead to miscommunication, which in turn can lead to distrust and a crisis of legitimation between the various institutions and the constituencies that they serve.

While regulatory agencies may need to be cautious in their approach to new food technologies, the public has no such restrictions. As long as exit and voice options are available, consumers and citizens have the opportunity to speak with their voices or their pocketbooks to force change within the marketplace (Hirschman 1970). Advocates of genetically modified foods are currently facing this situation since a large number of consumers are asking for nongenetically modified foods, at least according to various vocal actors (e.g., *Farm Journal,* <www.farmjournal.com>; *Natural Law Party News,* <www.natural-law.org>). This is not to say that all consumers are fearful of biotechnology and genetically modified foods. In a review of surveys conducted in the United States, it was found that, while consumers typically had a low level of knowledge concerning biotechnology, their understanding of the promises of technology in general led them to have an overall positive outlook toward this technology (Hamstra 1998). One caveat to consumer acceptance was that, while scientists could be trusted, they needed to be held accountable and operate under stringent regulatory standards. What we need to consider at this point is whether or not scientists, industry interests, regulatory agencies, and reporters are playing from the same page. If they are not, then public opinion will begin to turn against the authoritative and legitimate power centres of society.

Conclusion: Disaffection and Delegitimation

What is clear is that at the beginning of the twenty-first century a profound revolution is occurring that is radically transforming food production through the control and modification of the genome (Rifkin 1998). These new biotechnologies represent not only profound changes in the way that agriculture is done but also the social production of nature (Goodman and Redclift 1990). Humans now have the power to design nature at very fundamental levels for purposes that they deem necessary: health, profit, power (Busch et al. 1991). Such changes not only create new plants, animals, and food products but also generate important and sweeping social, economic, political, and ethical impacts.

The source of the controversy surrounding this revolution is that the radical transformations of food production were occurring "while we were asleep." Without much announcement and fanfare, biotechnology companies, in cooperation with university-based scientists, and with the approval of federal and state regulatory agencies, were genetically engineering crops and introducing bioengineered hormones and starter cultures in milk, beef, and cheese production.

Exit and Voice: But Where Is Loyalty?

What has surprised many in government and industry – the systemworld – is the swiftness with which consumers and national and international actors have found alternatives to the dominance of GMOs in the market. GMOs are competing against conventional and organic foods. Nations and food processors are now asking for a segregation of GMO and non-GMO food products. Two major baby food producers, Heinz and Gerber, will not accept genetically engineered ingredients (Pollack 2000). These corporations are trying to preserve brand name loyalty and trust in their products' quality and safety. Warren Belasco (1993) found that US consumers are typically loyal to brand names for many of their food purchases, leaving GMO producers and processors competing in a saturated market where brand names rule.

As Albert Hirschman (1970) has shown, loyalty is an extremely powerful force in the marketplace. GMOs are actually a catalyst for switching to organic foods; organic regulations do not allow modified foods in any form. As major firms adopted GMOs, the exit option was to shift to organic and other alternative foods. In short, there is little overt consumer demand for GMOs. Without a unified message articulating the beneficial qualities of these products, GMOs will have a long, uphill battle to win a significant portion of the produce marketplace. On the other hand, GMOs are being purchased extensively by large processed food manufacturers who already enjoy brand loyalty among a large fraction of consumers. These activities offer opportunities for activist opponents to attack the corporation and thereby to diminish consumer loyalty; on 23 March 2000, Greenpeace activists unfurled on Kellogg's corporate headquarters a large banner depicting "Tony the Tiger" as a Frankenstein monster.

A critical driver in this legitimation crisis is risk perception. Underlying such perceptions are knowledge, control, communication, and cultural understandings (Beck 1999). On the one hand, there is some evidence that, as individuals believe they have more knowledge of a hazard, they gain more control over their exposure to it (Frewer et al. 1996, 1998). The response, even though quite late, of biotechnology corporations and governmental agencies to develop an informational campaign on GMOs is an acknowledgment of this principle. But both pro- and anti-GMO groups must recognize that control of technology is not generally an important construct or focus of public concern. Only when there is a specific and definite threat or a hazardous event will people develop such concerns and experience a sense of worry (Sjoberg 1997). This was evident in the rise of public concern about food safety in Europe after the outbreaks of mad cow disease and foot-and-mouth disease and the contamination of cooking oil with dioxin. In the United States, the first alarm was raised over the introduction of an unapproved, potentially allergenic, genetically altered variety of corn and the difficulty of removing it from the food chain. Subsequent events included the contamination of native Mexican maize with GMO pollen, the unrealized mixing of GMO soybeans with non-GMO corn, and the apparent release for commercial slaughter of genetically modified hogs.

Yet the basis of public perception of risk is rooted in cultural meanings and understanding. When there are risks that challenge long-held worldviews or ideologies based on deeply held beliefs and long-established ways and habits of everyday life that are supported by strong social relationships, people are more likely to view their "realities" as being fragile (Wildavsky and Dake 1990). Rather than facilitating individual knowledge or personal characteristics, the larger frameworks of culture provide understanding and predictions of risk perception and risk taking.

This is the lesson to be learned from the legitimation crisis of food safety and GMOs. Unless new technological and scientific innovations successfully link with existing cultural frameworks, a crisis can emerge that challenges the perceived safety and acceptability of a new product or technology. And if they are challenged, people will voice opposition and may look for an exit option. These voices have opened up the possibility of democratic discourse on the future of the biotech revolution. The discourse may challenge the existing practices in the agrifood system and in the food safety system. But with increased knowledge and greater efforts to develop evidence of the safety of GMOs, consumers may accept bioengineering and its products as legitimate elements of their food lifeworld. The future of bioengineering of food and food products will be in jeopardy until the food producers and processors, the multinational corporations, and the food safety regulatory system learn this lesson.

Acknowledgments
The authors wish to thank the Michigan Agricultural Experiment Station, MSU Extension, and the National Food Safety and Toxicology Center at Michigan State University for their support for this research. An earlier version of this chapter was presented at the Hawaii Sociological Association Annual Meetings: Legacies of Marginalized People at the Dawn of the New Millennium, on 5 February 2000.

References

Albrecht, Don E. 1997. "The Changing Structure of U.S. Agriculture: Dualism Out, Industrialism In." *Rural Sociology* 62: 474-90.

Altheide, David L., and Robert P. Snow. 1991. *Media Worlds in the Postjournalism Era*. Hawthorne: Aldine de Gruyther.

Bailey, Britt. 1998. *Against the Grain: Biotechnology and the Corporate Takeover of Your Food*. Boston: Common Courage Press.

Beck, Ulrich. 1999. *World Risk Society*. Cambridge, UK: Polity.

Belasco, Warren J. 1993. *Appetite for Change*. Ithaca: Cornell University Press.

Bell, David, and Gill Valentine. 1997. *Consuming Geographies*. New York: Routledge.

Berger, Peter L. 1969. *The Sacred Canopy*. New York: Anchor Books.

Berger, Peter L., and Thomas Luckmann. 1971. *The Social Construction of Reality: A Treatise on the Sociology of Knowledge*. London: Penguin.

Bijman, J. 1999. "Life Science Companies: Can They Combine Seeds, Agrochemicals, and Pharmaceuticals?" *Biotechnology and Development Monitor* 40: 14-19.

Bridgstock, Martin. 1982. "A Sociological Approach to Fraud in Science." *Australian and New Zealand Journal of Sociology* 18: 364-83.

Brown, Patrick. 2000. "The Promise of Plant Biotechnology: The Threat of Genetically Modified Organisms." On-line at <http://www.psrast.org/promplantbiot.htm>.

Busch, Lawrence. 1997. "Grades and Standards in the Social Construction of Safe Food." Paper presented at the conference on The Social Construction of Safe Food, Norwegian Technical University, Trondheim.

Busch, Lawrence, William B. Lacy, Jeffrey Burkhardt, and Laura R. Lacy. 1991. *Plants, Power, and Profit: Social, Economic, and Ethical Consequences of the New Biotechnologies*. Cambridge, UK: Blackwell.

Caplan, Pat. 1997. *Food, Health, and Identity*. London: Routledge.

Chomsky, Noam, and Marcus G. Raskin. 1987. "Exchanges of Reconstructive Knowledge." In Marcus G. Raskin and Herbert J. Bernstein, eds., *New Ways of Knowing,* 104-56. Totowa: Rowman and Littlefield.

Clarke, Adele, and Theresa Montini. 1993. "The Many Faces of RU486: Tales of Situated Knowledges and Technological Contestations." *Science, Technology, and Human Values* 18: 42-78.

Collins, Harry, and Trevor Pinch. 1993. *The Golem*. New York: Cambridge University Press.

Committee to Ensure Safe Food from Production to Consumption. 1998. *Ensuring Safe Food: From Production to Consumption*. Washington: National Academy Press.

Corporate Watch. 2000. *Functional Foods: Good for Monsanto's Health*. On-line at <http://www.corporatewatch.org/publications/GEBriefings/funcfoods.html>.

Counihan, Carole, and Penny Van Esterik. 1997. *Food and Culture*. New York: Routledge.

Dawkins, Kristin. 1997. *Gene Wars: The Politics of Biotechnology*. New York: Seven Stories Press.

De Visser, A.J.C., E.H. Nijhuis, J.D. van Elsas, and T.A. Dueck. 2000. *Crops of Uncertain Nature? Controversies and Knowledge Gaps Concerning Genetically Modified Crops: An Inventory*. On-line at <http://www.mindfully.org/GE/Knowledge-Gaps-Greenpeace-Wageningen.htm>.

Dearing, James W., and Everett M. Rogers. 1996. *Agenda-Setting*. Thousand Oaks, CA: Sage.
DiMaggio, Paul. 1990. "Cultural Aspects of Economic Action and Organization." In Roger Friedland and A.F. Robertson, eds., *Beyond the Marketplace,* 113-36. New York: Walter de Gruyter.
–. 1994. "Culture and Economy." In Neil J. Smelser and Richard Swedberg, eds., *The Handbook of Economic Sociology,* 27-57. Princeton: Princeton University Press.
Durant, John, Martin W. Bauer, and George Gaskell, eds. 1998. *Biotechnology in the Public Sphere*. London, UK: Science Museum.
Etzkowitz, Henry. 1996. "Conflicts of Interest and Commitment in Academic Science in the United States." *Minerva* 34: 259-77.
Fox, Nicols. 1997. *Spoiled: The Dangerous Truth about a Foodchain Gone Haywire*. New York: BasicBooks.
Frewer, L.J., C. Howard, D. Hedderley, and R. Shepard. 1996. "What Determines Trust in Information about Food-Related Risks? Underlying Psychological Constructs." *Risk Analysis* 16: 473-86.
–. 1998. "Methodological Approaches to Assessing Risk Perceptions Associated with Food-Related Hazards." *Risk Analysis* 18: 95-102.
Gamson, William A. 1992. *Talking Politics*. New Amsterdam: Cambridge University Press.
Gaull, Gerald F. 1998. *Biotechnology Regulation in America and Europe Viewed in a Cultural Framework*. London: Institute of Economic Affairs.
GeneWatch UK. 2000. "Monsanto's 'Desperate' Propaganda Campaign Reaches Global Proportions." On-line at <http://www.genewatch.org/Press%20Releases/pr15.htm>.
Goldman, Karen A. 2000. "Bioengineered Food: Safety and Labeling." *Science* 290: 457-59.
Goodman, Michael, and Michael Redclift. 1990. *Refashioning Nature*. New York: Routledge.
Grocery Manufacturers of America. 2000. *GMA Survey Shows Americans Learning More about Biotechnology: Food Consumption Patterns Unchanged*. Washington: Grocery Manufacturers of America.
Gunter, Valerie J., and Craig K. Harris. 1998. "Noisy Winter: The DDT Controversy in the Years before *Silent Spring*." *Rural Sociology* 63: 179-98.
Habermas, Jürgen. 1973. *Legitimation Crisis*. Boston: Beacon.
–. 1987. *Lifeworld and System: A Critique of Functionalist Reason*. Vol. 2 of *The Theory of Communicative Action*. Boston: Beacon.
–. 1989. *On Society and Politics*. Boston: Beacon.
Hajer, Maarten A. 1995. *The Politics of Environmental Discourse: Ecological Modernization and the Policy Process*. New York: Clarendon Press.
Hamstra, Ir Anneke. 1998. *Public Opinion about Biotechnology: A Survey of Surveys*. The Hague: European Federation of Biotechnology, Task Group on Public Perceptions of Biotechnology.
Hazelrigg, Lawrence. 1989. *Claims of Knowledge*. Tallahassee: Florida State University Press.
Hendrickson, Mark K., and William D. Heffernan. 2002. "Opening Spaces through Relocalization: Locating Potential Resistance in the Weaknesses of the Global Food System." *Sociologia Ruralis* 42: 347-69.
Hesman, Tina. 2001. "Biotech Firms Need to Address Emotional Issues of Consumers." *St. Louis Post-Dispatch,* 20 February, C8.
Hirschman, Albert O. 1970. *Exit, Voice, and Loyalty: Responses to Decline in Firms, Organizations, and States*. Cambridge, MA: Harvard University Press.
Hogan, Michael J. 1997. "George Gallup and the Rhetoric of Scientific Democracy." *Communication Monographs* 64: 161-79.
International Food Information Council. 2001. "More US Consumers Expect Biotech Benefits: Mixed Feelings, but Not Major Concern over Labeling." On-line at <http://www.biotech-info.net/expecting_benefits.html>.
Ipsos-Reid Agri-Food Research Group. 2000a. "Awareness of Genetically Modified Foods Wide but Knowledge Inch Deep: 55% Admit They Know Little about the Issue." On-line at <http://www.ipsos-na.com/news/pressrelease.cfm?id=1040>.

–. 2000b. "Canadian Awareness and Perceptions of Genetically Modified Foods: Seven in Ten (68%) Would Be Less Likely to Buy a Food Product If They Knew It Contained Genetically Modified Ingredients." On-line at <http://www.ipsos-na.com/news/pressrelease.cfm?id=967>.

–. 2000c. "Significant Knowledge Gap in Debate over Modified Foods: Most Concerned about Health and Safety Risks." On-line at <http://www.ipsos-na.com/news/pressrelease.cfm?id=1039>.

Isserman, Andrew M. 2000. "Mobilizing a University for Important Social Science Research: Biotechnology at the University of Illinois." *American Behavioral Scientist* 44: 310-17.

Kamaldeen, Sophia, and Douglas A. Powell. 2000. *Public Perceptions of Biotechnology*. Guelph: University of Guelph Department of Plant Agriculture Food Safety Network.

Kleinman, Daniel L., and Jack Kloppenburg Jr. 1991. "Aiming for the Discursive High Ground: Monsanto and the Biotechnology Controversy." *Sociological Forum* 6: 427-47.

Kloppenburg, Jack Jr. 1991. "Social Theory and the De/Reconstruction of Agricultural Science: Local Knowledge for an Alternative Agriculture." *Rural Sociology* 56: 519-48.

Krimsky, Sheldon, James G. Ennis, and Robert Weissman. 1991. "Academic-Corporate Ties in Biotechnology: A Quantitative Study." *Science, Technology, and Human Values* 16: 275-87.

Larsen, Christian, and William White. 1913. *Dairy Technology*. New York: Wiley.

Latour, Bruno, and Steve Woolgar. 1979. *Laboratory Life*. Princeton: Princeton University Press.

Levidow, Les. 2000. *"Sound Science" as Ideology*. Cambridge, MA: Harvard University Center for International Development, Science Technology and Innovation Program.

Levidow, Les, and Susan Carr. 2000. "Unsound Science? Transatlantic Regulatory Disputes over GM Crops." *International Journal of Biotechnology* 2: 257-73.

Liebeskind, Julia P., Amalya L. Oliver, Lynne Zucker, and Marilynn Brewer. 1996. "Social Networks, Learning, and Flexibility: Sourcing Scientific Knowledge in New Biotechnology Firms." *Organization Science* 7: 428-43.

Littlejohns, Michael. 2000. "Scientists Cautioned about Genetically Modified Food." Earth Times News Service. On-line at <http://www.gene.ch/gentech/2000/msg00073.html>.

McCarthy, Thomas. 1984. *The Critical Theory of Jürgen Habermas*. Cambridge, UK: Polity Press.

Massieu, Trigo, and Cristina Yolanda. 1991. "Pesticides and Biotechnology: Multinational Power." *Sociologica* 6: 129-50.

Monsanto UK. 2000. "Statement from Monsanto in Response to GeneWatch Press Release 04/09/00." London: Monsanto UK.

Nelkin, Dorothy. 1996. "An Uneasy Relationship: The Tensions between Medicine and the Media." *Lancet* 347: 1600-3.

No More Scares. 2000. "Wild Anti-GMO Claims Continue." On-line at <http://www.nomorescares.com>.

Pickering, Andrew. 1995. *The Mangle of Practice*. Chicago: University of Chicago Press.

Pollack, Andrew. 2000. "Talks on Biotech Food Today in Montreal Will See U.S. Isolated." *New York Times,* 24 January, A10.

Pollan, Michael. 2001. "Great Yellow Hype." *New York Times Magazine,* 4 March. On-line at <http://www.biotech-info.net/yellow_hype.html>.

Powell, Douglas, and William Leiss. 1997. *Mad Cows and Mother's Milk*. Montreal: McGill-Queen's University Press.

Priest, Susanna Hornig. 2000. "US Public Opinion Divided over Biotechnology?" *Nature Biotechnology* 18: 939-42.

Priest, Susanna Hornig, and Allen W. Gillespie. 2000. "Seeds of Discontent: Expert Opinion, Mass Media Messages, and the Public Image of Agricultural Biotechnology." *Science and Engineering Ethics* 6: 529-39.

Rifkin, Jeremy. 1998. *The Biotech Revolution*. New York: Jeremy P. Tarcher-Putnam.

Salter, Liora. 1988. *Mandated Science: Science and Scientists in the Making of Standards*. Boston: Kluwer Academic Publishers.

Schmidt, Ronald H., and Gary E. Rodrick. 2002. *Food Safety Handbook*. Hoboken, NJ: Wiley.

Seife, Charles. 2001. "Papal Science: Science and Religion Advance Together at Pontifical Academy." *Science* 291: 1472-74.

Shortbridge, Barbara G., and James R. Shortbridge. 1998. *A Taste of American Place*. Lanham: Rowman and Littlefield.

Sjoberg, Lennart. 1997. "Worry and Risk Perception." *Risk Analysis* 18: 85-93.

Swidler, Ann. 1986. "Culture in Action: Symbols and Strategies." *American Sociological Review* 51: 273-86.

Teitel, Martin, and Kimberly A. Wilson. 1999. *Genetically Engineered Food: Changing the Nature of Nature*. Rochester, VT: Park Street Press.

Ten Eyck, Toby A. 1999. "Shaping a Food Safety Debate: Control Efforts of Newspaper Reporters and Sources in the Food Irradiation Debate." *Science Communication* 20: 426-47.

–. 2000. "Interpersonal and Mass Communication: Matters of Trust and Control." *Current Research in Social Psychology* 5: 206-24.

Ten Eyck, Toby A., Paul B. Thompson, and Susanna H. Priest. 2001. "Biotechnology in the United States: Mad or Moral Science?" In George Gaskell and Martin W. Bauer, eds., *Biotechnology 1996-2000,* 307-18. London: Science Museum.

Ten Eyck, Toby A., and Melissa Williment. 2004. "The More Things Change ... : Milk Pasteurization, Food Irradiation, and Biotechnology in the *New York Times*." *Social Science Journal* 41: 29-41.

Thompson, John B. 1995. *The Media and Modernity*. Stanford: Stanford University Press.

Thompson, Paul B. 1997. *Food Biotechnology in Ethical Perspective*. New York: Blackie.

Turner, James S. 1970. *The Chemical Feast*. Washington, DC: Center for Study of Responsive Law.

Uchtmann, Donald L., and Gerald C. Nelson. 2000. "U.S. Regulatory Oversight of Agricultural and Food-Related Biotechnology." *American Behavioral Scientist* 44: 350-77.

Vaidyanathan, A. 2000. "Research for Agriculture: Some Current Issues." *Economic and Political Weekly* 35 (33): 2919-21.

Vogt, Donna U., and Mickey Parish. 1999. "Food Biotechnology in the United States: Science, Regulation, and Issues." *Congressional Research Service Reports,* on-line at <http://www.ncseonline.org/NLE/CRS/abstract.cfm?NLEid=16553>.

Weber, Max. 1968. *Economy and Society*. New Amsterdam: Bedminster.

Wildavsky, Aaron, and Karl Dake. 1990. "Theories of Risk Perception: Who Fears What and Why?" *Daedalus* 119: 41-60.

Wright, Susan. 2001. "Legitimating Genetic Engineering." *Dissent* 48: 62-69.

Wuthnow, Robert. 1987. *Meaning and Moral Order*. Berkeley: University of California Press.

Znaniecki, Florian. 1952. *Cultural Sciences*. Urbana: University of Illinois Press.

5
Genetically Modified Foods in Norway: A Consumer Perspective

Margareta Wandel

> In the opinion of the members of the panel there is no need for genetically modified foods in Norway today, because the availability, choice of selection and quality of conventional foods are sufficient. There are too many risk factors tied to the process of genetic modification in relation to foods.
>
> The most important effort in the area is probably to contribute to a well-informed, conscious and critical general public in relation to genetically modified foods. Such an effort may make valuable contributions to political work, policy formulation, control and decision making, and for the consumers. (Lay Panel Conference 1996, 8)

Two lay panel conferences about genetically modified (GM) foods were held in Norway, one in October 1996 and one in November 2000. The theoretical framework and practical outline of these conferences are discussed below, in the section titled "Theoretical and Methodological Background." Here it is sufficient to say that the purpose was to create a dialogue between laypeople and experts on this difficult theme. This approach can be viewed as a way to explore complex and divisive issues and encourage public debate in the interest of the public good, which is an important distinction of a cohesive society, as pointed out by Michael Mehta in Chapter 1.

The sixteen lay panel members were the driving forces in the conferences, and they wrote the final documents from the conferences, whereas the experts functioned as a consultative body to provide substantive knowledge. The second lay panel conference consisted of the same individuals as the first (with the exception of one member). The conclusion from this conference was to recommend a moratorium, which included a prohibition against the cultivation, import, and sale of GM foods and GM fodder in Norway, with the exception of cultivation of genetically modified organisms (GMOs) in field trials (Lay Panel Conference 2000). Participants in the 2000 conference recommended that the moratorium should not be removed until there is increased knowledge about long-term effects on the environment and human health, better coordination of laws and regulations, and increased efforts directed toward inspection, control, and traceability of GM foods.

The restrained attitude toward GM foods and GM technology shown by the panel is much in line with results from larger surveys. A representative survey of the Norwegian population showed that approximately 60 percent believed that gene technology in the production of foods should not be encouraged (Heggem 1999). This attitude is also shown, even more strongly, if the question focuses on the respondents in their role as consumers. In another nation-wide survey, respondents were asked about aspects of importance for them when considering the purchase of food. One of the most important aspects was that the food had not been produced with GMOs. This got the highest rating of all aspects listed, which also included, among other things, nutritional value, environmentally sound production, animal welfare, and additives (Torjusen 2001).

The reluctance of Norwegian consumers is also evident in more specific questions regarding GM foods. When asked about the likelihood of buying genetically modified tomatoes with a prolonged shelf life, or potatoes modified for disease resistance, 68 percent and 62 percent respectively answered that they would surely or most probably not buy such products if they appeared in the Norwegian marketplace (Heggem 1999). Genetically modified animal foods evoked the strongest opposition, with between 75 percent and 77 percent reporting that they would not buy salmon or pork made particularly lean by GM modification. The fact that many Norwegian consumers prefer lean pork did not help the acceptance rate.

Although European consumers in general seem to be more strongly opposed to this new technology, compared with consumers in Canada or the United States, Norwegian opposition seems to be rather exceptional within the European context. A survey of consumer opinions about food risks in some European countries showed that 77 percent of Norwegian consumers, compared with 46 percent from Britain and 51 percent from Belgium, responded that they did not want GM foods (Berg 2000).

The strong and rather coherent picture shown by Norwegians is even more curious given that no foods, or ingredients produced through genetic modification, are presently approved for sale in Norway (Norwegian Food Authority 2004). Tests have shown that the Norwegian market is not completely free from such foods since imported foods may contain GM components without being labelled and often without the importer knowing this. This means that consumers have not knowingly been exposed to these foods in Norway, and most of them do not have direct experience with the consumption of such foods.

There has been an ongoing, and at times very lively, public debate about GM foods in Norway. Many researchers and other experts have tried to convince the public that there is no more risk associated with the consumption of GM foods than with conventionally produced foods. Much of the information given by experts relies on the premise that the techniques involved are just a development and refinement of breeding techniques that have evolved through thousands of years and that the main difference is that breeders have more control over genetic modification than conventional breeding. There is often an implicit assumption that the choice of breeding technique is in essence the business of the breeders, the food industry, and the food authority. The involvement of and reaction by consumers are often not understood or accepted as legitimate.

An intriguing question arises. Why do Norwegians have such strong opinions regarding foods with which they have no experience? There are many reasons for the reluctance to accept this new technology. In the following section, I will discuss some factors that may explain Norwegian consumers' reactions and positions on the question of GM foods. I will take as a point of departure some of the statements made at the two lay panel conferences. First, after presenting a theoretical background, I will direct attention to some ethical issues and particularly focus on "nature" and the "natural." Second, I will consider the issue of safety, risk, and trust from a consumer perspective, including risk-benefit valuations. And third, I will discuss the issue of consumer choice and labelling.

Theoretical and Methodological Background

Theories about Food Risk and Public Concern

The lay panel focused on risk factors and the importance of a conscious and critical general public. In attempting to clarify our understanding of the transformations of modernity, different authors have developed their ideas of what they call reflexive processes and their consequences (Bauman 1991; Beck 1992; Giddens 1991). These bear different relationships to the concept of risk and imply different characteristics of scientific knowledge or expert systems, which are agreed to be central to those transformation processes (Wynne 1996). The work by German sociologist Ulrich Beck is especially relevant to the issues taken up in this chapter.

In the book *Risk Society: Towards a New Modernity*, Beck (1992) hypothesizes that concern about food-related risks has increased, even though there has not been an actual increase in risk exposure. Beck classifies the risks of earlier times as "perceptible to the senses," such as the risk of food shortage or spoiled food, whereas many of the risks of today "are localized in the sphere of physical and chemical formulas" (21).

Beck discusses food risks in light of a general heightened risk awareness and skepticism related to science and technological progress. He postulates that the first half of the twentieth century was characterized by an unbroken trust in science and progress, and argues that, as a result of reflexive modernization, the end of that century focused on the negative consequences, "side effects," and risks produced by science and technology. Furthermore, he maintains that the participants in knowledge production have changed. Different organizations and interest groups claim to partake in the interpretation of scientific data. The harmonizing formula that "technical progress equals social progress" that prevailed in industrial society now comes under pressure (Beck 1992, 201). In this way, Beck argues, science has lost its monopoly of the truth. The authority of science is reduced, and the consumer, facing several alternative "truths," is obliged to choose which one to believe.

Other social scientists have related consumer emphasis on food risks to the special position of food among consumer commodities and to the widespread trade in food across countries and continents. Claude Fischler (1980) postulated that consumers have a problem with food in that the path from production to the table ("field to fork") has become long, complex, and less transparent. Ben Fine and Ellen Leopold (1993, 149) maintain that "the intimacy between food and the body requires absolute trust in the probity of supplies and suppliers" and that "this trust is often threatened by the growing awareness of deliberate intervention designed to deceive the consumer." The underlying theme is that consumers, to an increasing degree, are losing control over the origin and quality of the food they consume and that this loss may affect their view of food in the long term. The loss of control has to be replaced with trust that the food has been produced in a sound way, has not been adulterated, and is safe for consumption.

These observations were made before GM foods became a consumer issue in Norway and before the debates really took off in most EU countries. However, it can be argued that consumers' perceived loss of control of the origin and quality of food is particularly germane in the case of GM foods. People are concerned that GM techniques in food production are being used to make quick changes toward something very different from traditional foods, and they do not know the consequences in terms of safety or quality in general. The title of the report from the first conference, *Quick Salmon and TechnoBurgers: Report from the Lay Panel Conference of 1996,* reflects this concern.[1] Analogous to the theories discussed above, the report from the lay panel conference focused on long-term effects for health and well-being of both humans and the environment as well as consumer perceptions of food in general.

A Consumer Perspective

The statements of lay panel members reflect their roles as both consumers and citizens. This is in agreement with the theoretically oriented literature on public participation in technological assessments. Anneke Hamstra (1997, 53) writes that "the general public has two, not completely separable, essential roles with regard to new technologies: a role as consumers of the applications of the new technology, and a role as citizens in democratic society."

As Hamstra indirectly points out, the distinction between interests of consumers and citizens is neither easy nor distinct. Consumers represent the end point in the food chain from production to consumption. In this way, an issue tends to become a consumer issue the moment a product enters the marketplace. However, aspects that consumers find important in their evaluation of the end product may have arisen at an earlier stage of the food chain. In some instances, people may use their power as consumers to select commodities or services according to their concerns. This is highly visible with regard to questions about environmental issues (e.g., organic food consumption), ethical issues (demand for warranty that child labour has not been used), and political issues (boycotts of certain countries or products) (Dulsrud 1992; Torjusen et al. 2001). In other issues, consumers may not have a choice that can enable them to express concern through consumption, and if consumers are to have an impact they must participate in decisions at an earlier stage in the food chain. This is a broader definition of "consumer" than what emerges in classical economic consumer theory, resting as it does on the image of consumers as utility maximizing, trying to find the combination of products and services that will give them (as individuals) the most satisfaction (Munthe 1982).

This broader definition of consumer is also salient in the discussion of consumer rights. Such rights were first launched by US president John F. Kennedy in the 1960s and are the basis of a set of rights developed subsequently in many Western European countries. Of special interest in connection with GM foods is the right to health and safety, the right to information, the right to choose, and the right to a clean environment (IOCU 1993). In Norway, official consumer policy includes a consumer's right to food that is safe for consumption (White Paper 40 1998-99).

If GM foods become available in Norway, then Norwegians will get them as consumers. Consumer concern about these foods may reflect individual interests in terms of health, safety, the environment, or other ideological or community-oriented issues.

A Democratic Approach to Technology Decisions

In the book *Ecological Politics in an Age of Risk,* Beck (1995) discussed technological decision making in the light of democratization. He described early capitalism as an entrepreneurial paradise where industry could begin projects without submitting itself to controls and agreements. Then came an era of state intervention that included public consultation and the foundation of laws and regulations. Today even this intervention is no longer sufficient. Beck maintained that, even when these arrangements are negotiated and signed, company management feels exposed to further conflict, resistance, public denunciation, and suspicion. He saw this as an "expression of a more developed democracy, where an expanded civic consciousness refuses to be excluded from participation without a fight – in making decisions that intrude upon our lives more palpably and hazardously than those susceptible to parliamentary measures" (11). He also emphasized the need for new rules for consultation and decision making comparable to the changes that have occurred in the relations of production.

Simon Joss and John Durant (1995) also saw the discussion of public participation in science and technology as part of a larger ongoing debate about the interrelationships between science and society. These authors pointed out that this debate is sustained by an awareness that the place of science in society has become problematic. In this regard, their work is in line with the hypotheses put forward by Beck (1992, 1995). They maintain that, while modern medical sciences have great public support, there are other areas, such as some energy production technologies, chemical technologies, and genetic engineering, where public enthusiasm for the benefits of new applications is tempered by ethical, social, and physical dimensions. Such concern underlies the continuing search for new ways of negotiating science and technology policy in the public domain.

Janus Hansen (2000) discusses two main democratization strategies with respect to technology. These strategies can be viewed as practical consequences of different views of society. They are built on conceptions of conflict vis-à-vis participation, which to a certain degree is consensus oriented. The "conflict strategy" is focused on differences and stimulates confrontation. Hansen uses protest groups as an example. Their demand for democratic influence is based on perceptions of themselves as self-organized groups, with roots in the society as a whole. They seek to influence the public debate through conflicts and scandals and to address problems in the political process.

On the other end of the scale, there is "participation strategy," which is built on a broad call for citizens who are influenced by the decisions at stake. Such strategies may be built on various grounds. In the light of new constructivist approaches to science, these strategies are often justified on the ground that there is a need for a rationality other than the strict technical rationality provided by experts. It is thereby hoped that such a procedure can lead to a more socially optimal rationality and strengthen the democratic component in technological and administrative decisions. Thus, it is the implicit normative aspect in expert evaluations that is challenged and, ideally, replaced with democratization.

Both of these strategies are viewed as contributions to the democratization process by facilitating collaboration between citizens and the established system; they are not a form of "guardianship." The conflict strategy can be said to influence established institutions indirectly and from the outside, whereas the participatory strategy involves joint work and the taking on of responsibility by laypeople. Hansen (2000) emphasizes that these two strategies should be viewed as analytical tools. In practice, the actors on the technology scene may pick elements from both strategies.

Public Participation in Technology Assessment

Technology assessment is one contribution to the negotiation of science and technology policy. In recent years, different forms of technology assessment have been institutionalized in many countries. According to Jon Fixdal (1997), there is a general agreement as to the concept of technology assessment. It entails a systematic assessment of the advantages and disadvantages of a technology, either in the design stage or later when the technology is already in use. Technology assessments are meant to be voices in the decision-making process.

Fixdal (1997) divides technology assessments into three distinctly different approaches. Incidentally, they also describe a development toward increased democratization. The *instrumental model* uses different groups of experts (also those who are directly involved in the technology under analysis) to identify and assess problems raised by a certain technology. The assessments are viewed as a means to increase the likelihood that politicians may foresee the impact of different technological developments on social processes and thereby be better equipped to control them. The *elitist model* tries to establish neutral judge panels that include scientists from fields other than those directly involved in the issues at stake. With this model, scientific knowledge is the justification whenever there is disagreement about the consequences of a technology or the direction of a development path.

Both the instrumental and the elitist models have been criticized extensively on, among other things, the ground of democratic deficiency (Fixdal 1997). With regard to the first, it has been pointed out that experts often are uncertain and disagree about the impacts of various technologies on social conditions. Thus, their advice to politicians may be highly divergent and deficient. The second has been criticized on the ground that science and technology problems are viewed in isolation from the political responsibility of democratic institutions and transferred to groups of scientists who are not organized to be responsive to democratic principles. It has also been pointed out that technology assessments not only aid in decision making but also ought to have an important function within a democracy as a means to raise awareness and form consensus, thus directly referring to the desirability of a cohesive society.

As a result of the critique raised about the two former models, the third model, which Fixdal (1997) calls the *participatory model,* has been developed, and in it the public has been allocated a central role. There is also an increased focus on normative aspects and on defining the scope of the problem itself. Whereas the problems were given to panels in the first two models, in this model discussions about how to define, limit, and formulate the extent of the problem are seen as an important part of the recommendations that the panel puts forward. One of the practical approaches within this model is the so-called consensus, or lay panel, conference.

Referring to the two strategies in the former section, we can say that, even though lay panel conferences normally function as instruments within the participatory strategy, they can also be viewed as confrontations between two broad "interest groups": experts and nonexperts (Cronberg 1995). The nonexpert lay panel members may represent a variety of interests, but they have one thing in common: they have no professional interest at stake in the issue. The opposite is often true of the expert panel.

Historically, the consensus conference is a development of the so-called consensus development conference (CDC), which has been used in the United States from the 1970s on to assess the use and safety of medical technologies. When the Danish Teknologirådet (Technology Council) in the 1980s tried to develop a model for technology assessment that could promote communication between experts, politicians, and the public, it adopted the CDC model and developed it further along the lines of a participatory model.

According to the Danish model, a conference proceeds in several steps (Fixdal 1997). First, organizers choose the theme for the conference. Next, advertisements are placed in newspapers to get participants for the lay panel, and between fourteen and sixteen individuals are selected. There is a preparatory part with seminars for panel members. A conference is then held, at which both experts and the lay panel participate, that lasts for about three days and ends up with a report from the lay panel. The choice of the word *consensus* stems from the idea that a unanimous statement is thought to be of more interest to politicians than a number of single statements. The designers of this model have also explicitly pronounced that they want to see how far one can come when decisions are to be made on the basis of unity (Klüver 1995).

The Norwegian Lay Panel Conferences

It was the Danish consensus conference model that was used for Norwegian lay panel conferences. According to Fixdal (1997), the Norwegian conferences were not explicitly designed as a result of the developments from the instrumental to the participatory model. However, the instrumental model incorporates all of the main elements of the participatory model. It is based on the recognition of the need for a broad public debate about the impact of technology developments on social conditions, that laypeople play a central role in this definition and assessment of the problem area, and that specification of themes is endowed with great importance.

The recruitment process for participation in the lay panel included advertisements in ten Norwegian newspapers. Of the 380 individuals who replied expressing an interest in participation, 16 were chosen with relation to age, sex, regional affiliation, and occupation to make the panel as broad as possible. Care was taken to make sure that members were not professionally connected to the issue at stake. The panel had two weekend seminars before the conference in order to learn to work together, to get an introduction to the theme, and to work out questions to be answered during the conference. Each conference lasted three and a half days. On the first day, twelve to fifteen experts gave lectures on a large variety of aspects connected to the questions worked out by the lay panel. The second day was dedicated to questions from the lay panel and experts. During the remainder of the conference, lay panel members worked on the report. It was expected to answer questions generated by the lay panel prior to the conference.

Many of the lay panel members had, in the four years that had elapsed between the two conferences, become engaged in activities related to the issue of biotechnology in a variety of ways. Some of them had entered studies that had given them increased knowledge in biotechnology. Thus, in the second conference, they were no longer simply laypeople.

Experiences from Consensus Conferences

The important question on how successful the consensus conferences have been in shaping technology is not easy to answer. There has not yet been an evaluation of the political impact of the two conferences cited here, which have been the only ones carried out in Norway. Analyzing the experience of several conferences in Denmark, Tarja Cronberg (1995) writes that the consensus conference is seldom integrated into a concrete decision-making process and that there is no direct access to officials. The final report, media coverage, and public access to the conference provide indirect access and means of coercion. Thus, even though the organizers may be state or parastate institutions, the democratic effect is largely indirect, creating public debates on the topic selected for scrutiny.

Cronberg (1995) maintains that two types of shaping processes have taken place in Denmark. First, she asserts that, subsequent to the conferences, Parliament has enacted legislation to control technology. She writes that, more often than not, the final reports have been critical of a new technology, and she points to experts' enthusiasm for new technological advances and lay panels' introduction of a "more balancing, less enthusiastic voice" (128). In this way, politicians have had an alternative channel for information and opinion creation rather than relying only on experts. Second, Cronberg emphasizes a less obvious shaping process: the influence on experts. The lay panel agenda, questions, and interpretations often tended to take the experts by surprise and gave them a new frame of reference from which to judge the problem. Another reaction registered is that experts, known to disagree, become more favourable to each other's views when faced with questions from the lay panel.

It has been emphasized that there are a number of limitations to consensus conferences as the only means of involving the public in technological development (Cronberg 1995; Fixdal 1997). First, each conference deals only with one topic, and represents only a "one-shot" participation at a certain point in time, and therefore does not contribute with a permanent voice in the shaping of technology in a socially responsible fashion. Second, the shaping of technology negotiated in a consensus conference is located at the "use" interface of technological change. New properties of existing or potential technologies are seldom the result of a consensus conference. This critique also involves accusations of consensus conferences as social engineering and doubts regarding the lay participants' degree of representativeness. Fixdal (1997) writes that there were great differences in the way that politicians responded to the statements from Danish conferences. Some paid great attention because laypeople wrote them, and others pointed out that the panel constituted a small and unrepresentative part of the population and therefore should not get attention.

It seems right to conclude that the lay panel conference by no means represents a final solution to the problem of democracy in technological decisions, but it may be a first step toward increased openness and public participation. This approach utilizes social capital by actively building forums that encourage public debate on difficult and controversial questions like those posed by the development and subsequent commercialization of genetically modified foods.

GM Foods: Nature and the Environment

Nature as an Expression of Authenticity

> In the opinion of the members of the panel, the GM techniques are clearly different from traditional breeding techniques. This represents something new. We feel the need to stop and come to a decision of where we stand in this matter ... Respect for life and nature is part of our identity. What will happen with us if we use GM techniques in food production in an uncritical manner? Are we in danger of being truncated human beings? (Lay Panel Conference 1996, 35)

This quotation focuses on Norwegians' view of themselves and their relationship to nature. The members of the panel are worried that acceptance of GM foods may endanger this relationship. They have not accepted the information from biotechnologists and plant breeders that GM techniques are continuations of traditional plant breeding. They emphasize that this is something distinctively different. The statement above reflects the view that the use of these techniques will impinge on something central to the Norwegian identity: namely, respect for nature. They recommend that society stop and consider what these changes will bring about.

Norwegian researchers have also taken up the question of the place of GM foods in ideological concepts of "nature" and the "natural." In a study of quality perceptions by actors in different parts of the food chain, Runar Døving and Marianne Lien (2000) write that people in general conceive of vegetables as having come into existence by nature itself. The seed has a central position in the mythological consciousness as something that is inherently "natural" in the Platonic sense. Even if we no longer adhere to Plato's conception of the natural, it has influenced our thinking about an object's inherent naturalness. These ideas have persisted even though new breeding technologies have been in use for a long time. The authors argue that, with the advent of GM techniques in breeding, the idea of the seed as something inherently "natural" is being challenged.

Nature has been a fundamental concept in Norway since the national Romanticism movement in the 1850s, where wild nature was considered as the perfect reflection of the original Norwegian soul (Amilien 2003). This idea is also reflected by the finding that many foods that have a strong connotation of nature in the Norwegian ethos also have a strong position in the Norwegian food culture. Virginie Amilien also describes how Norwegian restaurants use "nature" and "survival" to justify the authenticity of the food offered.

The concept of "natural" as opposed to "artificial" is also central in consumer judgments of what is good or bad. In a qualitative study of consumer opinions and knowledge about health, environmental, and ethical aspects of food, participants were found to dichotomize the properties of food that were important for them (Bugge 1995; Wandel 1997). One such dichotomy was natural versus artificial. Others were fat versus lean and heavy versus light. Foods that were classified as natural were by definition considered better than those classified as artificial.

The respondents in that study had been informed about food and health from different sources, and this was a way to live with this information in everyday life. In the analysis, this system was compared to how people who subscribe to the humoral system of medicine often classify foods in dichotomies, for example hot/cold or yin/yang (Wandel 1997). However, whereas the humoral system is seen as dynamic, which means that the needs for foods of different classes may change according to the state of the body (Logan 1977; Wandel et al. 1984), the classification in this study seems to be more static. Thus, consumers do not easily reclassify foods that have been designated as artificial.

The strong opposition of Norwegian consumers to GM foods may reflect that these have already been classified as artificial. The way people think about GM foods is also reflected by the way these foods are discussed in public. Laypeople often use concepts such as "gene manipulation," "gene fiddling," or "messing around with the genes," whereas "gene modification" or "genetic engineering" are used by professionals working in the area. In the public debate, experts try to convince people that GM techniques are just continuations of conventional breeding techniques as described above, whereas at the grassroots level they are thought and spoken of as something "artificial" and "alien." I will argue that the static properties with which the described classification system is endowed in Norway have implications for consumer views of these foods. In the future, it may make it difficult for change to come about.

GM Foods and a Healthy Environment

> There seems to be a general agreement among experts that there is very little knowledge regarding the effect on the environment of the use of GM plants. (Lay Panel Conference 2000, 3)

GM techniques can be used for the advancement of environmentally sound production – but they can simultaneously be a threat to the environment. The lay panel in the 2000 conference made this observation. However, in the context of consumer concern about GM foods, the focus is often on the threat to the environment. The lay panel, as well as environmental organizations and consumer organizations, have focused on concerns about genetic contamination of non-GM plants, the spread of genes coding for antibiotic resistance to micro-organisms, the eradication of environmentally friendly insects, and field trials that are too short to predict long-term effects (Hurtardo 2000; Lay Panel Conference 2000). The scope of this chapter is not to discuss the probability of risk to the environment but to point out that such fears are common, not only in small groups, but also in the general population.

Environmental consciousness has developed gradually since the 1960s, fronted by what have been called environmental movements. Different in composition from country to country, these movements have shared what Jamison et al. (1990) call a particular set of knowledge interests. One of them refers to cosmology – that is, the more or less taken-for-granted frameworks by which human beings impose meaning on the world, such as an ecological worldview stressing the systematic interconnection of natural and social processes. Human beings, seen as part of a system, have no right to make changes for shortsighted economic profit at the expense of other life forms and the integrity of ecosystems. Another knowledge interest refers to the use of technology, for example the criticisms of particular destructive technologies, and to specification of alternatives. In most countries, the chemical industries and their polluting products and processes were the first main object on the agenda. Many other issues, among them biotechnology and in particular genetic engineering, followed.

Organic farming has taken up many of the knowledge claims cited above. Among its main goals are the following: "To base production on a holistic view, which includes ecological, economic and social aspects, in a local as well as global perspective," "to manage the natural resources in such a way that harmful environmental effects are avoided," and "to maintain the genetic diversity of species" (Debio 1996, 3). The practices of organic farming, which are now regulated by law both in Norway and in the European Union, prohibit the use of GMOs along with a range of other substances, such as chemical pesticides, industrial fertilizers, and synthetic growth enhancers, in the production of food.

It is likely that Norwegian consumers' strong emphasis on nature and the natural also reflects their views of the impacts of GMOs on the environment. However, this emphasis has not resulted in a particularly high interest in organic foods. In fact, the consumption of organic foods is particularly low in Norway compared with other Northern European countries (Berg 2000; Wandel and Bugge 1994). This fact may partly be explained by the price difference between organically grown and conventionally grown produce, which is very large in Norway. Poor availability and insufficient information are also considered important barriers for consumption of organic food (Torjusen, Nyberg, and Wandel 1999).

Food can be categorized into distinctly different classifications according to the technology used to produce it: organic, conventional, and GM. Even though organically grown foods represent an alternative that should ensure that GM techniques have not been used, and often represent a shorter and better-defined distribution path (from farmer to consumer), these foods have a very low market share in Norway. With regard to the interface between conventional and GM foods, results from research trials in four Nordic countries, including Norway, indicated that consumers preferred conventional to GM foods, even though these were presented with statements claiming that GM foods were produced in ways beneficial to the environment (Grunert et al. 2000).

It is likely that the view of GM foods as distinctly different from traditionally produced foods is key for consumers. This view was articulated by the lay panel conference of 2000. It also focused on the lack of scientific data and knowledge as a basis for its recommendations and positions in this matter. Furthermore, people's view of the environment and nature as a whole, and their place in this view, may be important for how they conceive the threats that the production of GM foods may pose.

Food Safety, Risk, and Trust from a Consumer Perspective

Food Safety and Risk

> We cannot, based on research results that are presently available, exclude the possibility of serious health consequences ... The expert statements in the conference of 2000 do not indicate that the uncertainties tied to environment and health have changed substantially during the four years that have elapsed since the last conference. (Lay Panel Conference 2000, 3, 6)

The above statements focus on risk to human health from the use of GM foods. The report also points out that the lay panel is aware that conventionally produced foods do not offer absolute safety, but it argues that this is not a good enough reason for an unconditional approval of GM foods (Lay Panel Conference 2000). Furthermore, the members convey that they feel a responsibility to act since there are potential risks associated with GM foods.

The Eurobarometer has indicated that consumers in Northern European countries (with the exception of Finland) are more concerned about the risks of genetic modification of food compared with those in Southern Europe. Frewer et al. (2000) postulate that this may be because consumers in Northern Europe are more "risk averse." The authors use this concept to suggest that Northern Europeans base decisions about food consumption on avoiding risks, whereas they claim that consumers in Southern Europe may be more concerned about potential impacts on food quality. However, the authors also warn that care must be taken with interpretations about consumer attitudes and acceptance, since some of the concerns shown in Northern Europe are shared by consumers in some Southern European countries.

Survey data from Norway have shown that the use of gene technology in food production tops the list of areas that respondents found most risky in biotechnology. The other areas were the production of medicines and vaccines and genetic testing (Heggem 1999). Research on the attitudes toward and preferences for specific foods has also shown that consumers consider that health and safety are endangered by GM foods. In a study of consumer perceptions of a number of foods, conventionally produced products were associated mainly with safety, good health, or good taste, whereas any kind of GM application was associated with uncertainty and deemed less healthy, along with a host of more specific negative associations. This in spite of the fact that the GM foods examined were described as having different kinds of benefits, such as good taste, high quality, low calorie content, et cetera. However, in none of the cases were the benefits able to compensate for the negative associations of GM foods in general or the more specific risks that the respondents attributed to them (Grunert et al. 2000). Thus, gene technology in food production seems to be a focal point for reflexivity, as Beck (1992) defines it in relation to the "risk society."

It is interesting to note that the results from the lay panel conferences do not indicate that the increased knowledge gained at and between the two conferences made the members substantially more in favour of gene technology. These results are in agreement with experiences from an educational effort among farmers in Norway (Almås 1994). Increased knowledge about biotechnology has been proposed as one of several means to make people more in favour of it. However, empirical research has shown varying results. Data from the United States have shown that efforts to increase awareness and knowledge top the list of factors that positively influence attitudes toward biotechnology (Hoban 1998). However, working with data from Europe, a research group found that, wherever there had been a wide public debate about the risks and benefits of this technology, providing further information was more likely to prime attitudes already held than to persuade the public of a particular view (Frewer et al. 2000). More data from different groups of people are needed in order to find out if there are specific cultural differences with regard to this relationship.

Another question concerns use of the precautionary principle in the assessment of risk. Statements by the lay panel reflect that it is using the precautionary principle as a basis for making judgments. This principle was launched so that governments could act (with prohibition) to protect health and the environment when there are threats of serious irreversible damage, especially in cases where the potential consequences of the threat have not yet been fully assessed by science.

This principle has been included in many international conventions (for GMOs, the key function was agreed to in the Cartagena Protocol on Biosafety, article 10, 2000) and widely discussed by experts. Interpretation of this principle may range from requirements to formulate risk models based on current evidence to more unspecified claims that there are too many uncertainties with a particular technique, practice, or product, so that measures of prohibition are justified (Economic and Social Committee 2000; NOU 2000, 29).

The lay panel's statement that "we cannot exclude the possibility of serious health consequences" and that there is no indication "that the uncertainties ... have changed substantially" show that the panel has used this principle in a less restrictive way than what is proposed in some expert documents, for example the one from the EU Economic and Social Committee (2000). This discrepancy probably reflects a different rationality in technical assessments than what experts provide, which was discussed by Hansen (2000) with regard to the participatory strategy.

Before the second lay panel conference, there was an evaluation of safety for human health carried out by an expert committee, appointed by the

Norwegian government (NOU 2000). The majority of members came to the conclusion that there is no research published according to general scientific principles that shows that GM foods are unsafe for human health. The members agreed that it is impossible to make judgments that apply to all GM foods, since each GM food represents a unique set of problems. Thus, each new food has to be judged separately on a case-by-case basis. A salient feature that was pointed out in this as well as other evaluations is the lack of research on health implications of GM foods. Different potential health implications of GM foods were discussed in relation to the precautionary principle, and there was disagreement as to the use of this principle with regard to GM foods.

Data from Norwegian consumer studies have shown that consumers in general are interested in the health and safety aspects of food and, to varying degrees, pay attention to these aspects in their choices of food (Bugge 1995; Wandel 1997; Wandel and Fagerli 1999). However, they also articulated that they found it difficult to have to pay so much attention to health aspects. They pointed out that there is much to learn about their own food. Many wanted easy directions to follow so that they could then relax and enjoy the food.

It would be very difficult for the general consumer to follow judgments about GM foods in detail. Even when consumers are very interested in the issue, there is a limit to their involvement, particularly since many other risks related to food also attract consumer attention. In this situation, the issue of trust becomes particularly important. However, European consumer trust in the food system and food authorities has been challenged due to several food scandals, including the outbreak of mad cow disease (Jansson 1998; Smith 1991). Thus, European consumers' reactions to GM foods have to be seen against the trustworthiness or lack of trustworthiness with which they endow the experts and producers, food industry, and food authorities.

Risk: Benefit Considerations

> In our valuation of health aspects, we see it as important to not only judge the health risks as a basis for a moratorium, but also to evaluate these risks in relation to the benefits which may be achieved with the new technology. (Lay Panel Conference 2000, 3)

The clear statement that the panel sees the risks in light of the benefits which the new technology may provide is also reflected in the main conclusions cited in the introduction. In that statement, the panel members considered the benefits to be too small when related to the risks because the Norwegian food supply at present is sufficient in terms of availability, choice of selection, and quality.

The risk/benefit way of thinking that these statements reflect also comes to the surface in many other ways, both in the public debates about this new technology and in official Norwegian documents. The Norwegian Gene Technology Law (2 April 1993, number 38) states that all GMOs should be evaluated in terms of the benefits for society, ethical justifiability, and whether or not they contribute to sustainable development.

Norwegian population surveys have shown consistently that respondents are more positive regarding gene technology when it is put to use in medicine rather than in food production (Heggem 1999). People are less concerned about the risks, and they see more benefits when the technology is used in this way (except in some cases, such as introducing human genes into animals for the production of transplant organs). These results are consistent with the work of Joss and Durant (1995). It is likely that such results also reflect considerations about the relations between risk and benefit.

In discussions of risk/benefit, there is always a question regarding who will reap the benefits from a new technology. It is logical to assume that those who reap the most benefits will also be more positive about the new technology and may be more willing to accept risks.

The statements from the lay panel conference of 2000 focused on promises given by the biotechnologists in the first lay panel conference that this new technology would soon be of benefit to consumers. Panelists noted that *"the positive aspects of GM foods which we were promised 4 years ago ... have not yet been realized"* (Lay Panel Conference 2000, 3). With this statement, they referred to the fact that none of the foods approved for sale in the European Union, and none of the applications made to Norwegian food authorities, included modifications that would benefit consumers, for example in terms of quality or health promotion. The panel members were aware that such foods were in the development process, but they had not seen any evidence that this would be a substantial part of the new technology.

Researchers in other parts of Europe have also discussed consumer perceptions of GM foods with regard to perceived benefits. Using research from several European countries, Frewer et al. (2000, 33) write that "further increases in consumer negativity towards genetically modified foods appear to have arisen because of the order of entry of products into the market place. The European public perceived that the first genetically modified foods available were of benefit to industry rather than the consumer. Novel foods with direct and tangible consumer benefits are more acceptable than those from which only industry will benefit or profit."

Thus, the feeling among consumers that they must embrace some level of risk with no perceived benefits may have contributed to the reluctant attitude toward these foods.

Consumer Trust

> The disagreement between experts is still rather great – different conclusions are drawn from the research results that are presently available ... The members of the lay panel conference pointed out four years ago that ecologists seem to be more skeptical than biotechnologists about the effect on the environment/ecological systems. At that time we meant that it was natural to attach the greatest importance to the ecologists' judgement. Today this is even more clear to us. (Lay Panel Conference 2000, 4)

In relation to the description of the risk society made by Beck (1992), there is a postulation that there are many authorities on issues related to risk, that there is an increasing focus in society on differences in opinion among experts, and that people choose the authority they trust most. Trust in experts and other authorities can no longer be taken for granted. Adopting a stance analogous to this postulation, the reports from both lay panel conferences focus on the dissension between experts (health, environmental, and biotechnology experts) about the risks associated with GM technology. This dissension was also apparent among experts who expressed their views in the lay panel conferences (1996, 2000). Debate in the media has also revealed dissension among experts.

The pattern of trust – that is, what consumers find more or less trustworthy – is important for how they perceive issues of health and risk. Information from more trustworthy institutions will have more impact on consumers as actors in the food market. Also important are consumer perceptions of who is the driving force in the new technology, for what purpose, and with what outcome.

Since none of the GM foods on the market are grown in Norway, this raises the issue of Norwegian consumers' trust in foreign producers, and in international control and regulation systems, as well as whether or not strong international market forces pay due attention to consumer interests, such as health, safety, the environment, and ethical issues. It also has a solidarity aspect with relation to Norwegian farmers.

Studies have shown that a majority of Norwegian consumers prefer Norwegian-produced foods because they believe that these foods are the safest to use (Berg 2000; Wandel and Bugge 1994). These results are analogous to the notion that they trust Norwegian producers more than foreign producers. This higher trust in domestic food is not unique to Norwegian consumers and is found in studies including other Western European countries (Berg 2000). This is also consistent with Fischler (1980) that consumers may have problems regarding identification of food when the path from production to table becomes long and less transparent. According to Fine and Leopold (1993), this creates a special strain on the trust relations between consumers and other actors along the food chain.

With regard to solidarity, the report from the first lay panel conference (1996) states that "commercial production of GM foods will have economic and market related consequences. The use of this technology is presently expensive. High investment costs will favor the big food producers. Small producers who do not have capital enough to take part in the new technology will have trouble in keeping up with this new technology. Almost all Norwegian producers are small from a global point of view. High costs and patent rights may increase the risk of ending up with a strong concentration of production" (27).

The report touches on economic mechanisms and driving forces in the development of GM foods. It raises some doubts with regard to the functioning of market mechanisms. In the statement above, this doubt is directed toward the balance between large- and small-market actors and the consequences for Norwegian producers and production. The report also raises doubt whether global market forces and strong economic interests will pay enough attention to safety issues and whether government regulation and control are good enough to ensure the safety of these foods. The central issue here is related to consumer trust in the market mechanisms with regard to equity and safety of production and in the government authorities responsible for regulation and control.

Data from the Eurobarometer have shown that Norwegians are exceptional in that they have a great trust in government authorities and low trust in market mechanisms (Tufte and Ali 1998). Norwegians' high confidence with regard to food safety was confirmed in a comparative study of Norway, Belgium, and Great Britain (Berg 2004). Even though Norwegians in general trust government authorities, this does not mean that they accept all information given from these sources. Government institutions have informed the public about safety issues concerning GM foods in a manner that should reassure consumers. However, this information has not influenced the skeptical view that consumers have of these foods.

Furthermore, consumers prefer control to information when it comes to difficult food safety issues (Wandel 1997). In general, Norwegians seem to want government to take over the question of food safety as much as possible, and that focus is directed more toward control than information. The report from the second lay panel conference also recommended that efforts be directed toward building up a strong control and tracking system in Norway.

As mentioned previously, no GM foods have presently been approved for consumption in Norway. Thus, there has not been any conflict between the official position of the government and consumers' views of these foods.

Taken together, the data from consumer studies and the lay panel conferences indicate that these foods are viewed as "alien" in many respects. First, they are viewed as alien in that "manipulating" genes produces them. Second, they are produced and marketed by big international actors, none of which is Norwegian. And third, no GM foods are presently approved for consumption in Norway. Thus, these foods are not incorporated into existing trust relations, which are of great importance with regard to consumer valuation of food.

The Question of Labelling

> In the opinion of the panel, the products ought to be labeled as far as it is possible. We wish the consumers to be able to choose their own food. (Lay Panel Conference 1996, 34)

The conclusions about labelling of GM foods from the first conference are clear and consistent with the view that consumers should have the right to choose what they want to eat. At the second conference, the recommendations made do not include labelling since they suggest a moratorium that acts as a general prohibition against such foods.

Consumers need information in order to make good choices about the food that they consume. As mentioned earlier, this is also a central consumer right. Genetic modification, like many other production-related aspects of food, represents a special information challenge. Consumers may make up their minds about the desirability of GM foods by reading books or through the media. Nonetheless, to make the choice in the store they need labelling that can tell them whether a particular food has been produced through genetic modification or not.

However, there are several dilemmas regarding labelling. First, there is a general dilemma concerning the amount of information that could and should accompany a product. Consumers want to know as much as possible about their food. Consumer studies have shown that they want to know the contents of ingredients, nutrients, and additives, and that they are interested in knowing about the cultivation process, animal care, and country of origin (Wandel 1997). On the other hand, the results showed that present labels are too complicated for many consumers. It is a challenge to be able to simplify the labels at the same time as new information, such as that concerning GMOs, is added. The challenge is especially large regarding mixed foods with many ingredients and multiple sourcing.

Second, there has been a discussion regarding which GM foods should be labelled. Some GM foods have been highly refined before they enter the market, such as oils and sugar. These foods contain only trace DNA or protein. They are termed "substantially equivalent" to conventional foods in the approval process (EU Commission 1997; EU Parliament 1997; NOU 2000). This concept refers to the contents of the final products. However, in view of the many different reasons that consumers may have for the reluctance to buy these products, and in view of the process itself as distinctly different from traditional cultivation, it is questionable whether this term has any meaning for consumers.

Consumer organizations want clear labels on all foods in which genetic engineering has been used. This demand comes in response to survey findings that European consumers do want labelling (Hurtardo 2000). Consumer organizations also take the view that labelling of GM foods should be comprehensive. This means that all foods in which genetic modification has been used should be labelled, even if there is no trace in the final product of modified DNA. This demand relates to the fact that consumers may have different reasons for wanting to know if the foods have been produced with the help of GM techniques, and these reasons may relate to more than the final product.

On the other end of the food chain, the food industry is worried that food labelling would scare consumers off through a process of stigmatization. Faced with consumer pressure, especially in Europe, Australia, and New Zealand, some of the leading GM food producers have said that they will support labelling, but only when the GM food is different from the conventional one – in other words, if GM material (protein or DNA) is present in the final product (Hurtardo 2000).

The Norwegian labelling regulations from 1997 require special labels for all products produced with GM techniques, even if there are no traces of DNA or protein in the final product (Norwegian Food Authority 2004). This will also be the case in the European Union when the new regulation 1830/2003 about traceability and labelling of genetically modified organisms is implemented in 2004 (Swedish Food Authority 2004). However, the United States has signalled that it views the European Union's restrictive politics regarding GM foods as a trade barrier and thus in conflict with the GATS treaty (Norwegian Food Authority 2004).

The debate about labelling is still very intense and difficult to resolve. Basically, it is a question of who has the power to decide what consumers ought to know in order to make informed choices in the food market. It is also apparent that the various actors discuss different issues, since some producers insist that consumers should only be concerned about the final product, whereas many consumers, as we have seen, are also concerned about the process by which the food has been produced.

If we go back to the many stated reasons that consumers may have for their reluctance to eat these foods, it is evident that, if GM material (or its products) has to be detectable in order to require labelling, this caters fairly well to health and safety reasons but not to religious and ethical reasons or to those connected to concern for the environment. Thus, this debate reflects the perceived legitimacy on the part of the GM food producers and legislation responsible for consumers and their different reasons for wanting to know more about the foods that they buy.

Conclusion

Communication regarding GM foods is particularly difficult because of different views of what is important to know, insufficient knowledge, and lack of published research on health and environmental issues with regard to these foods. Basically, many of the disagreements between different actors in the food chain regarding introduction of, information on, and labelling of GM foods are questions of who has the power to decide what are important issues in the case of gene technology and what consumers ought to know and react to. We do not yet know what effect, if any, the lay panel conferences have had on political decisions or on consumer trust in food and institutions responsible for food safety and quality.

At present, no GM foods have been approved for sale in Norway. The reluctance that a majority of consumers show indicates that these foods may meet a rather unfriendly atmosphere the day they appear on the market. However, it is impossible to predict precisely how consumers will react when they have choices in the market. Food safety is but one of the issues of concern for modern consumers when they make their choices in the food market.

On the one hand, if government-approved GM foods appear on the Norwegian market, they may be incorporated into products and by producers that are already known and trusted. Some people may extend their trust to these products and thus change their views of these foods. However, this chapter has shown that there are many factors that may work to preserve held positions in the case of GM foods, such as the focus on natural, ecological concerns, reflections on risk, and health matters.

The results from this chapter suggest different dimensions that may be important for how Norwegian consumers will view GM foods in the future. These dimensions include the degree to which research efforts have been directed toward environmental, health, and safety aspects of GM foods, thereby contributing to increased knowledge in aspects that are important for consumers; whether or not there are control and tracking systems that consumers can trust; whether or not consumers are able to see more benefits for themselves when buying these foods – that is, if production is geared more toward aspects important to consumers, such as food quality, nutritional content, and safety; and whether or not these methods will improve the environment.

Note

1 The panel has not given any explanation for the title of the report, *Quick Salmon and TechnoBurgers*. However, it is obvious that this title connotes two very sensitive aspects for laypeople with regard to GMOs. The first concerns genetic modification of animals, which, according to the report, is especially disagreeable to the panel. This view was also evident among the responders in national Norwegian surveys (Heggem 1999). The second relates to consumers' desire to know what the food contains. The name "technoburger" connotes that this food could be produced from anything.

References

Almås, R. 1994. "New Biotechnology and the Greening of Politics: Value Cleavages, Economic Interests, and Public Opinion in Risk Society." *Sociologisk Tidsskrift* 2 (1): 23-40.

Amilien, V. 2003. "A Taste of Authenticity: Nature and Tradition in Norwegian Urban Restaurants Context." In M. Hietala, ed., *The Landscape of Food,* 213-26. Helsinki: Suomen Kirjallisuuden Sura.

Bauman, Z. 1991. *Modernity and Ambivalence.* Cambridge, UK: Polity Press.

Beck, Ulrich. 1992. *Risk Society: Towards a New Modernity.* London: Sage Publications.

–. 1995. *Ecological Politics in an Age of Risk.* Cambridge, UK: Polity Press.

Berg, L. 2000. Tillit til mat i kugalskapens tid; en komparativ karlegging med fokus på forbrukertillit og matsikkerhet i Norge, England og Belgia (Trust in Food in the Age of Mad Cow Disease). Report 5-2000. Lysaker, Norway: National Institute for Consumer Research.

–. 2004. "Trust in Food in the Age of Mad Cow Disease: A Comparative Study of Consumers' Evaluation of Food Safety in Belgium, Britain, and Norway." *Appetite* 42: 21-32.

Bugge, A. 1995. *Mat til begjær og besvær* (Health, Environmental, and Ethical Aspects of Food). Work Report 6-1995. Lysaker, Norway: National Institute for Consumer Research.

Cronberg, T. 1995. "Do Marginal Voices Shape Technology?" In S. Joss and J. Durant, eds., *Public Participation in Science: The Role of Consensus Conferences in Europe,* 125-32. London: Science Museum.

Debio. 1996. *Driftsregler før økologisk landbruksproduksjon* (Norwegian Certification Rules for Organic Farming). Bjørkelangen, Norway: Debio. On-line at <http://www.debio.no>.

Døving, R., and M. Lien. 2000. Myten "om den perfekte kålrot" (The Myth about the Perfect Rutabaga). *Norsk Antropologisk Tidskrift* 11: 108-27.

Dulsrud, A. 1992. "Boikott som forbrukerpolitisk virkemiddel" (Consumer Boycott as a Tool in Consumer Politics). In J.W. Bakke and M. Lien, eds., *Mellom nytte og nytelse,* 149-65. SIFO Work Report 9. Lysaker, Norway: National Institute for Consumer Research.

Economic and Social Committee. 2000. *Own-Initiative Opinion of the Economic and Social Committee on Use of the Precautionary Principle,* NAT/065. Brussels: European Union.

EU Commission. 1997. "Commission Recommendation of July 1997." *Annex to the Official Journal of the European Communities* (97/618/EC): L253/1-L253/35.

EU Parliament. 1997. *EC Regulation No 258/97 of 27 January 1997.* Brussels: European Parliament.

Fine, B., and E. Leopold. 1993. *The World of Consumption.* New York: Routledge.

Fischler, C. 1980. "Food Habits, Social Change, and the Nature/Culture Dilemma." *Social Science Information* 19: 937-53.

Fixdal, J. 1997. "Lekfolkskonferanser som teknologivurdering" (Lay Panel Conferences as Technology Assessment). In R. Søgnen, ed., *Verneverdig eksperiment? Evaluering av Lekfolkskonferansen om genmodifisert mat,* 75-111. NIFU Report 5/97. Oslo, Norway: Norwegian Institute for Studies of Research and Education.

Frewer, L., J. Scholderer, C. Downs, and L. Bredahl. 2000. *Communicating about the Risks and Benefits of Genetically Modified Foods: Effects of Different Information Strategies.* MAPP Working Paper 71. Aarhus, Denmark: Centre for Market Surveillance, Research, and Strategy for the Food Sector.

Giddens, A. 1991. *Modernity and Self-Identity: Self and Society in the Late Modern Age.* Cambridge, UK: Polity Press.

Grunert, K.G., L. Lähteenmäki, N.A. Nielsen, J.B. Poulsen, O. Ueland, and A. Åström. 2000. *Consumer Perception of Food Products Involving Genetic Modification: Results from a Qualitative Study in Four Nordic Countries*. MAPP Working Paper 72. Aarhus, Denmark: Centre for Market Surveillance, Research, and Strategy for the Food Sector.

Hamstra, A. 1997. "The Role of the Public in Instruments of Constructive Technology Assessment." In S. Joss and J. Durant, eds., *Public Participation in Science: The Role of Consensus Conferences in Europe,* 53-60. London: Science Museum.

Hansen, J. 2000. *Den teknologiske udfordring: En differentieringsteoretisk analyse af forholdet mellom demokrati og teknologi i et risikoperspektiv* (The Technological Challenge: A Differential Theoretical Analysis of the Relationship between Democracy and Technology in a Risk Perspective). Copenhagen: University of Copenhagen, Institute of Sociology.

Heggem, R. 1999. *Genteknologien sitt janusansikt: Ei studie av folk sine haldningar til genteknologi* (The Gene Technology Janus Face: A Study of People's Attitudes to Gene Technology). Report 7/99. Trondheim, Norway: Senter for Bygdeforskning.

Hoban, T.J. 1998. "International Acceptance of Agricultural Biotechnology." In R.W.F. Hardy and J.B. Segelken, eds., *Agricultural Biotechnology and Environmental Quality: Gene Escape and Pest Resistance,* 59-73. NABC Report 10. New York: National Agricultural Biotechnology Council.

Hurtardo, M.E. 2000. *GM Foods: The Facts and the Fiction*. London: Consumers International.

IOCU (International Organization of Consumers Unions). 1993. *Food Policy beyond 2000: A Policy on Food Issues in Industrialized Countries*. The Hague, Netherlands: International Organization of Consumers Unions.

Jamison, A., R. Eyerman, J. Cramer, and J. Læssøe. 1990. *The Making of the New Environmental Consciousness: A Comparative Study of the Environmental Movements in Sweden, Denmark, and the Netherlands*. Environment, Politics, and Society Series. Edinburgh: Edinburgh University Press.

Jansson, S. 1998. "Galna kor i moderna landskap: Frågor om mat och tillit i det nya Europa (Mad Cows in Modern Landscapes: Questions about Trust in the New Europe)." *Kulturella perspektiv* 3: 12-21.

Joss, S., and J. Durant. 1995. *Public Participation in Science: The Role of Consensus Conferences in Europe*. London: Science Museum.

Klüver, L. 1995. "Consensus Conferences at the Danish Board of Technology." In S. Joss and J. Durant, eds., *Public Participation in Science: The Role of Consensus Conferences in Europe*. London: Science Museum.

Lay Panel Conference. 1996. *Kvikklaks og teknoburger: Sluttrapport fra Lekfolkskonferansen om genmodifiert mat, 18-21 October 1996* (Quick Salmon and TechnoBurgers: Report from the Lay Panel Conference). De nasjonale forskningsetiske komitéer NEM, NENT, NESH. Oslo: Bioteknologinemnda.

–. 2000. *Et skritt mot mer kunnskap: Oppfølging av* Kvikklaks og teknoburger *fra 1996* (A Step toward More Knowledge: A Follow-Up of *Quick Salmon and TechnoBurgers* from 1996). Opp-følgingskonferansen om genmodifisert mat 2000, De nasjonale forskningsetiske komitéer NEM, NENT, NESH. Oslo: Bioteknologinemnda. On-line at <http://www.bion.no>.

Logan, M.H. 1977. "Anthropological Research on the Hot-Cold Theory of Diseases: Some Methodological Suggestions." *Medical Anthropology* 1 (4): 87-111.

Munthe, P. 1982. *Markedsøkonomi* (Market Economy). Oslo: Universitetsforlaget.

Norwegian Food Authority. 2004. *Genmodifisert mat: Gjeldende norske lover* (GMO Foods: Norwegian Laws in Effect). On-line at <http://matportalen.no/Saker/1063869784.82> and <http://matportalen.no/Saker/1074780653.27>.

NOU. 2000. *GMO-mat: Helsemessige konsekvenser ved bruk av genmodifiserte næringsmidler og næringsmiddelingredienser* (GMO Foods: Health Consequences of Genetically Modified Foods and Food Ingredients). Norges Offentlige utredninger 2000: 29. Oslo: Statens forvaltningstjeneste.

Smith, M.J. 1991. "From Policy Community to Issue Network: Salmonella in Eggs and the New Politics of Food." *Public Administration* 69: 235-55.

Swedish Food Authority. 2004. *Skärpta regler för genetiskt modifierade livsmedel inom EU* (More Restrictive Regulations for GMO Foods in the EU). Uppsala: Swedish Food Authority. Online at <http:www.slv.se/templatesSLV/SLV_NewsPage_7831.asp>.

Torjusen, H. 2001. *Forbruk av økologisk mat sett fra forbrukernes side* (Consumption of Organic Foods from a Consumer Perspective). SIFO oppdragsrapport 16. Lysaker, Norway: National Institute for Consumer Research.

Torjusen, H., G. Lieblein, M. Wandel, and C.A. Francis. 2001. "Food System Orientation and Quality Perception among Consumers and Producers of Organic Food in Hedmark County, Norway." *Food Quality and Preference* 12: 207-16.

Torjusen, H., A. Nyberg, and M. Wandel. 1999. *Økologisk produsert mat: Forbrukernes vurderinger og bruksmønster* (Organic Food: Consumers' Perceptions and Dietary Choices). Report 5. Lysaker, Norway: National Institute for Consumer Research.

Tufte, P.A., and A. Ali. 1998. *Innflytelse, tillit, og forbrukeratferd* (Influence, Trust, and Consumer Behaviour). SIFO Work Report 3. Lysaker, Norway: National Institute for Consumer Research.

Wandel, M. 1997. *Mat og helse: Forbrukeroppfatninger og strategier* (Food and Health: Consumer Opinions and Strategies). SIFO Report 5. Lysaker, Norway: National Institute for Consumer Research.

Wandel, M., and A. Bugge. 1994. *Til bords med forbrukerne: Forbrukernes ønsker og prioriteringer på matområdet i 90-årene* (Consumers, Food, and the Market: Consumer Valuations and Priorities in the Nineties). SIFO Report 2. Lysaker, Norway: National Institute for Consumer Research.

Wandel, M., and R.A. Fagerli. 1999. "Norwegians' Opinions of a Healthy Diet in Different Stages of Life." *Journal of Nutrition Education* 31: 339-46.

Wandel M., P. Gunawardene, A. Oshaug, and N. Wandel. 1984. "Heating and Cooling Foods in Relation to Food Habits in a Community in Southern Sri Lanka." *Ecology of Food and Nutrition* 14: 93-105.

White Paper no. 40. 1998-99. *Om forbrukerpolitikk og organisering av forbrukerapparatet* (Consumer Politics and the Organization of Consumer Affairs). Oslo: Statens forvaltningstjeneste (Ministry of Children and Family Affairs).

Wynne, B. 1996. "May the Sheep Safely Graze? A Reflexive View of the Expert-Lay Knowledge Divide." In S. Lash, B. Szerszynski, and B. Wynne, eds., *Risk, Environment, and Modernity: Towards a New Ecology,* 44-83. London: Sage Publications.

6
Commercializing Iceland: Biotechnology, Culture, and the Information Society

Kyle Eischen

Iceland Genetic Information and the Global Environment

In December of 1998, the Icelandic parliament passed legislation creating a genealogical database that covered the ancestry of a vast majority of Icelanders who have ever lived. The legislation explicitly included the right to link these genealogical records with medical records and tissue sample DNA that the Icelandic government meticulously kept on the population throughout the twentieth century. DeCode Genetics, an Icelandic genetics firm, was given sole custody of the development, control, and commercialization of the Iceland Genetic Database – or the Genotypes, Genealogy, Phenotypes, and Resources (GGPR) database, as it is officially known. DeCode will not use the database solely for in-house proprietary research or partnerships with government-sponsored medical research; rather, it will sell access to the database – and by extension the genetic information of the Icelandic population – to global pharmaceutical firms seeking to isolate genes and potential treatments more expeditiously.

The passing of the legislation and the resulting development of the database have caused intense debate within Iceland and deep concern throughout the world about the direction and impact of biotechnology on individuals and communities. Most of these debates have focused on very real and serious issues of privacy, competition, commercialization, and individual rights that challenge or extend existing local legal codes and social norms in fundamental ways. Equally important, and absent from current discussions generally, is understanding the development of the Iceland Genetic Database as part of new social, economic, and technological trends in the global environment. While concerns about the impact of "Decoding the Language of Life" (the deCode corporate tag line) are real, an equally significant issue is the new "social codes" being institutionalized and built upon and within Icelandic society.

From the global perspective, the developments in Iceland provide a way to outline how global economic, social, and technological trends shape and interconnect with local resources, needs, and policies. This link with global patterns in turn ties these local needs to other regions and institutions in the global environment, creating new relationships and dependencies distinct from past patterns. Iceland matters not only because of concerns over privacy or commercialization of genetic information on a regional level but also because the debate itself only exists when broader global trends impact in very real and powerful ways on specific regions and populations.

Social Cohesion, Institutions, and New Global Ties

For most of the last millennium, Iceland has been characterized by an inhospitable environment generating extreme geographic isolation with very traditional, limited linkages to the global economy through commodities exports (see Figure 6.1). Now, at the beginning of a new millennium, Iceland has become an increasingly central site of global biotechnology information and research, shifting radically the interest of the world in Iceland and the very isolation that signals its uniqueness. Understanding the nature of this shift and the development of the Genetic Database, why it can only exist in Iceland, and what social transformations stem from its development leads directly to detailing the forces that drive globalization and an information economy generally.

The genetic inheritance of Iceland, and its current economic and social value in the global economy, are derived directly from the intersection of a historically situated combination of natural selection, cultural homogeneity, and economic and geographic isolation. The very isolation that limited interactions with the global environment has also produced a deeply homogeneous, socially cohesive society. Over the course of a millennium, Icelandic society has developed cultural patterns and social institutions reflective of this isolation. What has fundamentally changed in the past decade is how the global environment views – and, more importantly, values – these cultural patterns and institutions. In the specific case of biotechnology, the social, cultural, genetic, and political cohesion of Iceland produces a unique "commodity" for an industry driven to find unique "information" sources and patterns.

National legacies, both institutional and cultural, that underpin Icelandic social cohesion played a key role in creating the genetic resources that the nation drew on to establish a credible and viable bio-information initiative as the global environment evolved to value exactly these resources. The dichotomy for Iceland is that biotechnology, by its unique production and organizational forms, creates new social tensions that challenge the very social norms that supported the creation of the initial genetic and social resources. Simply, the very process of building the database reinvents the relationships between doctor and patient, government and business, citizen

Figure 6.1

Overview of Iceland

- Population (2003): 290,570
- Capital: Reykjavík (population 2003: 113,387)
- Government: constitutional republic
- Founded: 874 CE
- Language: Icelandic and English
- Religion: 88.7% Evangelical Lutheran registered in State Lutheran Church
- Area: 103,000 km^2
- Wasteland: 60%
- Coastline: 4,970 km
- Fishing area: 758,000 km
- Average yearly temperature (Reykjavík): 41°F
- GDP per capita (2002): 29,619
- Key exports (2003):
  - Marine products: ~63%
  - Manufactured products (largely fishing products and aluminum): ~23%
- Health care spending (% of GDP 1998): 7%
- Health insurance (% of all national social protection expenses 1998): 37.3%

Sources: Statistics Iceland (2004); Central Bank of Iceland (2003).

and state, building on traditional norms and institutions while altering them in the process.

The building of the database intimately ties Iceland to the global economy and community in a way impossible to imagine, understand, or plan for without considering the overall global environment in which it has been developed. It also represents a fundamentally qualitative change in the interaction with the global environment that is hidden within economic statistics focusing on fish exports and tourism numbers. Information technologies such as biotechnology, both as processes and as products, contain and require embedded social knowledge and thus represent the construction of new social norms and institutions as local social knowledge is linked to the global economy. The construction of the database and a new biotechnology industry represent both challenges and opportunities for Iceland that come with the new patterns of work, competition, and global networks that embed local social structures within broader informational, economic, and social patterns.

The case of Iceland, as a new site of genetic information and biotechnology research, helps to elaborate exactly on the operation and impact of the new spatial and institutional structures that characterize the global environment. Linkages between the broader informational aspects of the global economy and local regions are becoming shaped within new norms, organizations, and institutions structured around the production, manipulation, and dissemination of information. The importance of defining and clarifying the unique informational aspects of industries such as biotechnology centres on understanding the social transformations that accompany and propel information-focused development. In other words, understanding biotechnology as an information-driven process, product, and industry helps to explain essential (and often seemingly contradictory) economic, political, and social features of the global environment.

It is exactly this combination of information technology and social transformation that is playing out in the case of Iceland. The establishment and development of the Iceland Genetic Database signals an "information society" in formation, with new beliefs and norms slowly institutionalized into new patterns and categories of power and profit, be they personal, economic, social, or political. These transformations underpin the central question of this volume, as Michael Mehta outlined in Chapter 1: "Will these new technologies unglue, or perhaps realign, the social fabric as we know it?" The rest of this chapter is devoted to understanding exactly this question for Iceland.

The Structure of the Informational Environment: Global, Networked, and Regional

The global environment is dynamic, structured on multiple spatial levels, and networked through multiple organizational forms and flows stemming from economic, technological, and social transformations simultaneously (Harvey 1990; Held et al. 1999). The combination is propelling the structuring of a new "information" or "networked" society characterized by the predominance of information as a central resource of both economic and social relationships (Castells 1996; Negroponte 1995; Webster 2002). Yet the specific operation and impact of the institutions, organizations, and norms structuring these new inchoate social relationships are only now becoming visible. By analyzing the informational aspect of these new social structures, particularly the operation of information technologies in specific regions (Castells 1989; Saxenian 1994), like biotechnology in Iceland, we can detail the impact of broader global patterns on local societies.

Understanding the impacts of informational and global trends on Iceland requires clarifying the structures that support such an environment both generally and locally. It also requires mapping the networks and flows through which the two aspects are interlinked and co-evolve within the global environment. Simply, the informational aspects of the global environment take specific process, product, and industry forms that dynamically interact with local institutions and cultures. These linkages are defined by new financial, social, and informational flows and networks that increasingly structure new patterns of power and profit between regions and the global environment. At the heart of these new relationships is information technology (IT).

IT has been clearly established as a central determinate of the direction and pace of economic and social change in the global environment (Castells 1996; Dicken 1998; Eischen 2000; Held et al. 1999). However, the role that IT plays in structuring new social interactions on both global and regional levels is often undefined or misunderstood. IT processes, products, and industries are the central means of generating, manipulating, and commercializing *information* in an information economy. This unique combination of information and production is the mechanism that links culture and economics in the global environment and in many ways signals what is "new" in the new economy and society. As such, IT is much more than just increasing means of transmitting information or increasing the speed and geographic spread of communication. Information technologies embody and require a very specific and distinct social basis that shapes their development as simultaneously a process, product, and industry (Agre 1995a, 1995b; Lessig 1999; Mitchell 1998). In this way, biotechnology, and other information technologies and industries, are much more than architectures for communication or tools of production in which an information society develops (Kranzberg 1985); they are ways to understand why information is valuable and how it is structured and valorized in the global economy (Eischen 2003a).

The complex nature and impact of IT in the global environment create a tension between theoretical perspectives that explore the impacts of information from either economic (the daily practice of surviving and producing the tools for survival) or social/cultural (new identities, social movements, and discourses) viewpoints. Such one-sided approaches, however, miss essential features of social change in an information-driven environment. On the economic side, it is easy to slip into economic and/or technological determinism, ignoring completely the socially embedded nature of material and scientific practices that structure how information is produced and embedded in specific institutions (Agre and Schuler 1997; Haraway 1997). On the social and cultural side, it is far too easy to forget that material practices matter, that the marshalling and institutionalization of culture and power through information historically have determined who eats, who receives medical care (and what kind), and who lives (Weber 1968; Williams 1958). Outlining the nature and impact of IT, as informationally and not technologically defined, provides a means to bring the economic and cultural together through an analysis of the global environment as informationally driven.

The development of the Iceland Genetic Database is a way to outline exactly this mixing of social, economic, and cultural factors that structure

Figure 6.2

Patterns operating in an informational environment

General patterns	Specific aspects
Informational processes, products, and industries	• Tacit, innovation-driven, and informational processes • Products embedded and defined by social knowledge • Global markets and regional, networked production for industries
Institutional mechanisms, norms, and flows structuring the informational environment	• Migration • Investment • Innovation networks • National and regional policy • Private-public partnerships • Entrepreneurial firms–global firm linkages
Local processes and structures interacting with the informational environment	• Regional economic and social networks • Unique local knowledge/information • Regional governance institutions and capacity • Unique social capital • Unique cultural practices • Regional educational and scientific institutions

an information society (see Figure 6.2). Thinking carefully about Iceland helps to clarify how the broader economic and social patterns in the global environment operate in a very specific and local way. The point here is to tease out some of the aspects of the "information society" through a consideration of biomedical policy in Iceland. A fuller understanding of what constitutes an information society, and by extension a detailed understanding of development in Iceland, require that we develop an analysis of both the local and the global, social and personal, and economic and cultural. The aim is to move beyond narrow considerations of privacy or ethics to understand that the experience of Iceland represents general trends in the global environment that are simultaneously extremely personal and local events.

Linking Information Technology to Social Practices and Institutions

The creation, manipulation, commercialization, and distribution of information in the global environment is facilitated by the basic algorithmic and digital patterns at the core of information technologies, enabling the manipulation of diverse information in its binary and defined forms (Eischen 2000). Software code, songs on Napster, basic DNA patterns, and the patterns on an integrated circuit all follow a basic binary pattern (ones and zeroes, open and closed, etc.) that can be manipulated using mathematically defined algorithmic formulas. This pattern shapes both the products produced and the organizational characteristics of IT industries that involve mapping and building these basic binary and algorithmic patterns. IT industries such as biotechnology are specifically organized to access, produce, transform, and commercialize specific and unique information that is at the core of the products they produce. Many of the most common features of the global environment (networked organizations, regional centres of innovation, skilled labour as a central resource, global immigrant networks) can be much more easily explained by viewing them from exactly such a perspective.

"Information"-focused economic initiatives are by definition specifically created to integrate and operate within global networks of production, immigration, innovation, and competition (Gordon 1994), the basic patterns that structure the generation and production of information products. However, because they are based on information and knowledge-based production practices, IT industries create organizations and institutions that operate in the global environment very differently from other globalized industries. Biotechnology and other IT industries need to capture or control the sources of unique information that drive innovation and competition globally. Accessing these sources of competition requires the networking of diverse and information-"rich" regions around the world, whether temporarily or permanently located in individuals, regions, institutions, or cultures. Biotechnology in this way is about the control of flows of information and the specific skills to handle that information, rather than the capturing of fixed, defined, and predetermined resources (Eischen 2003b).

Genetics is ideally suited to such an informationally focused environment, exactly because genes are one of the most fundamental means of storing, transmitting, and transforming information in existence (Armour 2000). This is enhanced because genetic information is structured around naturally occurring, if yet to be fully explained, algorithmic and digital patterns (Berlinksi 2000). Combined with other information technologies, particularly software and supercomputer technologies that easily blend with the binary and algorithmic patterns of biology, this basic genetic information has been transformed into a biotechnology industry (Fumento 2003; Rifkin 1998). The blending of biology and computer technologies has accelerated and transformed the cloning, antibody, genetic modification, and protein-engineering aspects of biological research and commercialization, and it has spawned entire new industries around biosensors, tissue engineering, DNA chips, and bio-informatics (Biotechnology Industry Organization 2001). Biotechnology has become simultaneously (1) a process of mapping genetic information, (2) a defined product derived from that mapping, and (3) an industry structured around replicating such processes and products.

The unique development processes, resources, and geographic organization of industries such as biotechnology are central to understanding global processes generally. This importance stems from the role of information as both an essential input and a final product of such industries and not necessarily the application of such knowledge by society in the form of a specific product or technology. In other words, while the creation of new genetically based drugs or foods may have significant, even frightening, social consequences, the greater immediate impact is the process through which such products are produced and how that process structures the end product itself. Information technologies, because of this informational characteristic, share this potential impact regardless of whether the final product is a drug, software, fruit, or animation (Eischen 2003a).

Aspects of an Informational Environment: Embedded Information and Culture

The unique aspects of information production actively structure organizational and institutional policies and practices on a local level. The advantages and challenges of information increasingly shape the strategic and organizational choices of governments, firms, and social actors, whether consciously or not (Lessig 1999) through the discovery, definition, and application of information. Thus, the building of a biotechnology industry in Iceland – by definition an information-focused process – has potential social impacts far beyond questions of privacy or competitive advantage.

Detailing the organizational and institutional structures that form to maximize informational production helps to signal the potential impact of industries such as biotechnology in Iceland. Following is a simplified typology of the key aspects and implications of information practices in the global environment (Eischen 2003a).

- The process is organized around the generation, definition, manipulation, and transmission of information into socially and economically applicable forms.
- Because it is socially structured and often determined, the production of such congealed knowledge will often take the form of craft-like (or creative or research-like) production systems where tacit (as opposed to explicit) knowledge is essential, peer or market review processes dominate, and production processes are nonrationalized.
- Skilled human resources, from multiple domains of knowledge, will be the central resource, with a weakly defined division of labour.
- Flexible networked organizations that are able to efficiently and rapidly manage the flows of information (both its creation and its applications), manage "information or knowledge workers," and implement new tools for its manipulation and dissemination will take precedence.

- Increasingly, value-added will be greater in the design or mapping of the algorithmic aspects of a process – that is, in the ability to define and model a process – than in its actual implementation, manufacture, or replication.
- The social knowledge and assumptions embedded in IT products will result in the institutionalization of norms prior to the application of the technologies in society.
- The social nature of domain knowledge – that is, the place – and the context-specific understanding of specific processes and practices ensure that regions and cultures will play a significant role in the development of IT.
- Production and firms will increasingly be globally defined, though product markets will remain fragmented.
- Monopolies will tend to occur through both the establishment of standards and the control of unique sources of information as firms push to control innovation and product cycles.

These general patterns are an essential feature in the global environment for transforming and linking local social knowledge and information – and thus culture and economy. In this way, the specific features of IT outlined above, particularly the need for locally specific domain knowledge, represent one of the essential mechanisms through which local information is linked to global economic and social patterns (Geertz 2000). Information technologies such as biotechnology are thus incredibly flexible and dynamic processes, not only comprising how information moves in the global environment but also requiring the embedding of such knowledge in the technologies themselves. As such, IT – whether in biotechnology, software, telecommunications, or multimedia – builds on existing institutions and norms while simultaneously creating new patterns that slowly intertwine with and transform existing social relations.

Exploring the implications for societies of these broader global patterns involves understanding the opportunities and challenges presented by the shift to information-focused processes, products, norms, and institutions. The Iceland Genetic Database needs to be placed within the general patterns that initiate and support information industries and production initially and then subsequently promote new norms and institutions around such industries and processes.

Iceland represents a confluence of national governmental policies, global trends, history, and cultural traits that has placed the region at the centre of a growing movement to isolate the genetic processes that cause disease and to patent key gene sequences. Focusing on biotechnology as a central economic resource and strategy is not just a simple choice of industry or policy for Iceland but also a transition to a new pattern of social and economic relationships – locally and globally. These new relationships detail how global forces interact with, use, and shape the personal lives and practices of Iceland's 290,000 people.

The Case of Iceland: Biotechnology, Culture, and Global Economics

Within the biotechnology industry, there is a global push to develop genetic databases and new processes of genetic screening that will enable the rapid development of new pharmaceutical products. The most "genetically pure" and documented populations offer the greatest benefits for such databases and research. Iceland is perhaps the most well-documented and pristine human genetic "biopreserve" within the new sectors of "bioprospecting" and "genomics" within biotechnology ("Norse Code" 1998; Shreeve 1999), but it is not by any means the only such site. Both Millennium Pharmaceuticals and Myriad Genetics have extracted DNA samples from "inbred" societies in Finland and Costa Rica as well as from the Mormon population in Utah (Billings 1999), and Autogen, an Australian biotech firm, has obtained exclusive rights to the genetic data of Tonga (Senituli and Boyes 2002). Other initiatives, initially sponsored by the National Institutes of Health and the Department of Energy (Marshall 1998) in the United States, sought to build vast archives of human genes for basic research. Some efforts have failed, as in China, where companies were accused of "biopiracy" and forbidden to transport genetic materials. Other private efforts, such as Axys Pharmaceuticals' research on asthma in the Atlantic island of Tristan da Cunha, have been seen as commercially viable.

Much of this push is tied to biotech research supported by large-scale "e-science," defined as "large-scale science carried out through distributed global collaborations enabled by networks, requiring access to very large data collections, very large-scale computing resources, and high-performance visualization" (UK Research Councils 2004). The push behind this evolution is trifold: (1) the failure of increased research spending to generate a corresponding increase in new drugs, pushing for maximizing returns, (2) the increasing sophistication, power, and declining cost of new technical tools for mapping, modelling, and developing new drugs, and (3) the resistance to and difficulties of using human or animal populations for direct genetic disease research. These developments have resulted in new ventures, such as deCode's Icelandic efforts, that serve as gateways between the local and the global, combining powerful computational models with population histories to isolate genetic sources of disease and model possible cures.

These general global industry patterns, however, have overlapped with unique cultural and historical factors to make Iceland a unique site of such research efforts. Historically, Iceland has been characterized by an intense and unplanned genetic homogeneity (Gudmundsson 2004; Kunzig 1998). Since its founding by ten thousand Vikings in 874 CE, Iceland has remained genetically homogeneous due to extreme geographic isolation compounded by natural disasters, including the plague in the 1400s (which killed two-thirds of the inhabitants), smallpox, and volcanic eruptions that caused widespread famine. Such disasters and the small founding population have caused population "bottlenecks" that have reduced the gene pool even further. As such, by historic accident, the Icelanders possess one of the "clearest bloodlines on the planet" (Moukheiber 1998), making them "ideal" subjects for genetic researchers seeking to isolate individual genes linked to specific diseases.

It is not, however, only this accidental homogeneity that makes Iceland so important. Four other factors linked to culture, governmental policy, and social networks have also been essential to shaping Iceland as a central site of global biotechnology research (Chadwick 1999).

1 Genealogy is practised widely. Most Icelanders can trace their ancestry back centuries, most to the year 1000, when Iceland converted to Christianity and priests began recording births and deaths. These genealogical records represent the basic raw material upon which the database is constructed. More importantly, while genealogy is an individual and private family activity, the institutional support and documentation provided by the local churches historically and more recently by the national health care system provide a source of genealogical information that can be accessed publicly. Through meticulous inputting of public records, the vast majority of the Icelandic population's genealogy can be re-created without the assistance or permission of individuals. The end result enables current individual genetic and medical information to be tied to family bloodlines that go back 1,000 years.
2 Iceland has a highly centralized, extensive, and effective public health care system that has been developed over the past 100 years. The first national pension program was established in 1890, and the first comprehensive social insurance legislation was passed in 1936. In 1990, primary health care was completely centralized under national control, with national public health insurance instituted in 1993. In 1999, the year of the debate over the database, over 87 percent of all medical expenses were paid by the national government, with over 47 percent of the national budget dedicated to health and social security expenses (Central Bank of Iceland 2003). The structure of the health care system raises several important aspects for the development of the database. First, the role of the national government as central provider of medical services and research establishes a social and institutional structure that fits well with the construction of a national genetic database. Centralized authority and control enable an extensive and economical "buy-in" of the new policy while minimizing the risks to either the government or the private sector. Second, the large proportion of the national budget dedicated to health

care provides strong incentives for alternative sources of cost control and revenue to be developed. The database potentially provides both through access to new drugs and the use of national information as a revenue source. And third, national government approval is essential to tie individual medical records to the existing genealogical information. Because of the historic role of the state in national health care, such medical files exist under government control for the majority of the population going back to 1915, when the centralized state-run medical and welfare system was initiated.

3 As part of the same national health and welfare system, tissue samples have been taken from a large portion of the population since the mid-1940s as part of constructing disease libraries. These samples provide detailed individual genetic information that can be combined with genealogy and medical records to link disease patterns with distinct genes. Importantly, since the majority of Icelanders who have ever lived have done so since 1940 (in part due to the increased health care provided by the national medical system), the tissue samples provide DNA information for the majority of individuals mapped in the genealogy of the database.

4 Key players in the promotion of Iceland as both a site and an information source of biotechnology research have been returning overseas Icelandic biotechnology researchers, chiefly Kari Stefansson and the other founding members of deCode Genetics in Reykjavík. These individuals are central actors within the social networks linking the specific resources of Iceland with the dominant trends in biotechnology and the global economy. It is through the flows of these researchers that specialized and unique domain knowledge (the aspects of the Icelandic medical and social structure, connections to key government officials and institutions) is matched with global economic and social institutions (venture capital, leading research institutes, global pharmaceutical firms, new economic models). The role that such flows and networks play in the global environment is a generalized pattern that is not unique to Iceland or biotechnology but has played a central role in both conceiving and pulling together the global and local resources to build the Iceland Genetic Database.

Developing Iceland as a Global Site of Genetics Research

The role of deCode in developing the genetic database is a clear example of how global migration, economic flows, and business networks shape and connect regions to the broader global environment. Kari Stefansson was an Icelandic emigrant to the United States, where he became a Harvard Medical School professor specializing in neurogenetics. He returned to Iceland specifically to found deCode Genetics Incorporated in 1997, supported by

$12 million of US venture capital funding and a group of US-trained Icelandic doctors. DeCode focused on two initial projects: (1) researching genetic-based disorders such as multiple sclerosis and essential tremor, and (2) digitalizing Iceland's genealogical records from the past 1,000 years. The two projects were complementary halves of an overall mission focused on using the unique nature of Iceland's population, history, and culture to track genetic diseases.

The first project was initiated using teams of physicians from around Iceland that would interact with patients and their relatives and collect blood samples. It is important to note that this initial pattern of research through a network of affiliated local doctors would come to be standard operating procedure at deCode for all future projects. Once collected, patient history and blood sample information were sent to deCode in encrypted code format based on a proprietary algorithm developed by the firm. At deCode, the DNA samples are genotyped (i.e., genetically profiled), matched with phenotype, then linked to the genealogical information of the second project in order to trace patterns and isolate genes. The first success came quickly. Through the use of automated sequencing machines from PerkinElmer, the gene that caused essential tremor was located in six months rather than the two years it would have taken in a nonhomogeneous population.

It is the second project, though really an integral part of the overall research described above, that has driven the most controversy. DeCode's proprietary genealogical database traces the ancestry of 75 percent of the 750,000 Icelanders who have ever lived in history. It is deCode's most valuable asset because it allows for tracing of disease along family lines over 1,000 years, enabling genetic traits to be more quickly identified, as in the case of essential tremor. The rapid identification of such genes is an invaluable asset for any company developing new drug therapies in an innovative, intensely competitive, intellectual-property-driven global market for new pharmaceuticals.

On 17 December 1998, the Icelandic parliament, the Althing, passed a legislative bill (with only one dissenting vote) that gave deCode not only proprietary control over the genealogical database but also the right to link these genealogical records with medical records and tissue sample DNA that the Icelandic government meticulously kept during the twentieth century. The bill gives deCode a de facto and de jure monopoly over Icelandic genealogical and genetic data. However, deCode will not use the database solely (or even mostly) for in-house proprietary research. DeCode will sell access to the database to global pharmaceutical firms seeking to isolate genes and potential treatments more rapidly, following the pattern established in the essential tremor trial by deCode itself (Henderson 1999).

Within two months of the bill's passage, the first such agreement was reached. DeCode signed a five-year, $200 million agreement with

Switzerland's F. Hoffmann-LaRoche that focused on the discovery of genes with alleles or mutations that predispose people to twelve different diseases or illnesses, including cardiovascular, neurological/psychiatric, and metabolic conditions. The $200 million includes equity investment, research and milestone payments, and royalties, with Hoffmann-LaRoche having exclusive rights to develop and commercialize all pharmaceuticals and diagnostics. The exception to the last is that deCode will control all antisense and gene therapy products. Any drug therapies developed from the use of the GGPR database will be freely available to the population of Iceland during the life of the patent (seventeen to twenty years).

Proponents of the legislation specifically and the deCode mission generally are primarily government based, being led by the prime minister's party and government bureaucrats centred on the health ministry's Medical Ethics Committee. The project is justified on both economic and medical grounds. The main benefits are said to be improved preventative medicine, improved cost management in the health care system, better understanding of the genetic basis of disease, improved health services management, and local economic development. Opponents include the Icelandic Medical Association (IMA) and the Mannvernd (the Association of Icelanders for Ethical Science) established in October 1998 specifically to oppose the passage of the legislation. Public opinion prior to the passage of the bill hovered around 60 percent in favour (Henderson 1999).

Three issues frame the debates around the legislation and continue to be central issues that have not been resolved with the passage of the bill (Masood 2000). First, privacy is central. Even though data will be encrypted, there is the clear possibility that identities, especially for rare conditions, will be revealed. Furthermore, the system is dynamic, with new patient information being added continuously, which clearly leaves open the possibility of privacy being violated. Interestingly, the debate centres more on social concerns of privacy than on denial of access to medical care based on the revelation of specific genetic diseases or markers. The reason is that patients feel assured of accountability through the national health care system but not in the private sector. The contradiction is that the database itself is effectively privatized, raising the question of whether the government can exercise sufficient control over the private sector, especially a global pharmaceutical firm outside its jurisdiction, to ensure patient privacy.

Second, trust between doctors and patients is stretched since patient consent to pass information on to deCode is not required. The legislation was structured around de facto consent, meaning that, if a patient does not opt out of the system, consent is assumed (Merz, McGee, and Sankar 2004). Furthermore, information once entered cannot be retrieved, even if the patient opts out at a future date. The dead, however, can never opt out, which implies that family patterns can be established even if an individual chooses not to participate. This is extremely significant given the medical and DNA samples that the government has included in the database. Even opting out of the system in the present does not ensure that a previous generation's genealogy, medical records, and DNA will not be used to construct an in-depth profile of a current patient.

Third, because deCode and its partners have been given an effective monopoly over the database, domestic researchers and competitors are essentially locked out of future research and, perhaps most importantly, oversight of deCode's practices. This monopoly on access limits both third-party accountability that is essential to government and public oversight and the effectiveness of the database itself. General trends within the information economy place increasing importance on the need for peer review, competition, and transparency to maximize the reliability, applicability, and monitoring of information products. This is even of more concern in the case of deCode, where the stated business model structures the firm as a portal to information and control, much more than an active research firm. Overall, the lack of access and oversight raises the serious question of whether medical research derived from the database will ever reflect the needs of the Icelandic population or merely reflect the priorities of third parties external to the local society.

Commercializing Culture: Iceland as a Sign of Things to Come

The case of Iceland's Genetic Database, and the controversies flowing from it, open the possibility of understanding not only the specific issues of privacy, trust, and competition (Greely 2000) but also how such issues are framed by broader economic, social, and technological trends in the global environment (Bowring 2003). Looking at the Iceland database project helps to clarify the multiple challenges and opportunities arising from an information-structured society. A traditional viewpoint would imply that Iceland is symbolic of what might occur in other countries, but that is not entirely accurate and misses a much deeper point. First, an Icelandic-type database could never happen elsewhere in the same way because the mix of history, culture, government policy, and economics is unique to Iceland. However, and this is the deeper point, it is this uniqueness that is a competitive advantage for Iceland in the current global environment exactly because it cannot be replicated. Simultaneously, the world benefits from this uniqueness through the linking of Iceland to global networks of research, information, capital, and migration. In other words, the structure of the global environment means that a database (or any information resource) does not need to be replicated everywhere, because essentially Iceland's database is accessible to the world through global firms, technologies, and markets.

This is exactly what has happened in practice since passing of the legislation. Through a series of global partnerships modelled after the initial

LaRoche alliance – including Wyeth, Merck, IBM, Emory School of Medicine, and Affymetrix, deCode has produced gene discoveries linked to stroke, osteoporosis, rheumatoid arthritis, asthma, and hypertension (deCode Genetics 2004). All are examples of the global and local links around which the database evolves. Most recently, scanning the genomes of 713 patients from the more than 8,000 heart attack patients listed in the database found a cluster of tiny variations in one gene linked to a twofold increase in the risk of heart attack. This led to a global research and development partnership with Bayer AG (Winslow 2004). DeCode and the database are performing exactly as intended, linking the intensely local with the demands of global markets.

Iceland's 290,000 people are a unique confluence of culture, geography, and knowledge that forms an ideal laboratory for such "bioprospecting" and "genomics." This distinct history intimately and uniquely ties them to broader global economic and social patterns. The genetic inheritance of Iceland, and its current economic and social value, are derived directly from the intersection of a historically situated combination of local culture (a combination of natural selection, cultural homogeneity, and economic and geographic isolation) and global trends (global networks of economics and technology). Traditional approaches to understanding this, either by communities or by governments, may not explain this interaction. For example, to say that selling 1,000 years of social and genetic knowledge for $800 per person,[1] including the dead, is not a good bargain does not capture the full magnitude of what is occurring. It may, in fact, be a very good deal, considering that for 1,000 years such information had no economic value outside Iceland and maybe very little inside it. No matter what the answer to this question is, there is a more fundamental question hidden within it: what has changed that has allowed such information to be valuable now, and how does that help to explain potential new social institutions and tensions?

Biology has always been an important source of both material and symbolic practice.[2] The specific global trends of "biomedicalization" and the building of a global "biomedical industrial complex" (Clarke and Olesen 1999) help to elaborate not only on the continued importance of biological models but also on the specific structure and impact of information as an industry in the global environment itself (Zucker, Darby, and Brewer 1994). The case of Iceland is an entry point into the blurring distinctions and the multiple combined meanings that an information society produces. While the creation of various genetic collections around the world is clearly exclusionary, excluding groups with too much genetic "noise" and alienating people from what is in some ways their most basic material inheritance, it is at the same moment intimately inclusionary. It requires, in fact, that those inside the chosen group all participate to yield the greatest results (the rationale for the de facto consent for the Iceland Genetic Database), and it includes them individually since each DNA molecule is considered to possess unique information. The most personal of information, DNA, becomes an input into a global biomedical industry that potentially links an individual Icelandic family with completely unrelated communities and individuals globally. In this way, the Genetic Database links the utmost local with the utmost global trends.

The very process of building the database reinvents the relationships between doctors and patients, states and firms, and publics and governments, building on traditional norms and institutions while altering them in the process. Each actor acts and reacts to both local and global challenges and opportunities, building new social relationships in the process. States privatize medical knowledge, weakening the link between the public and democratic regulation, while expanding the need for firms to work within the boundaries of the state's protection, while constantly seeking to justify to the public the benefits of such privatization. Weaknesses, in this case the geographic and cultural isolation of Iceland, become strengths that can be commodified as digital knowledge and information. Yet the very act of using those strengths alters the isolation of Iceland within a global world. By placing itself within networks of firms, people, biomedicalization, and knowledge, Iceland becomes intimately tied to global economic and social networks. DeCode, founded by an American-trained Icelandic doctor based at Harvard with US venture capital funds, digitizes and monopolizes the genetic history of Iceland and places it at the services of global genomics firms. Icelandic uniqueness becomes the basis for a universal understanding of gene traits through a global economic structure that requires exactly such unique information to operate.

In this way, the cultural and historical uniqueness of Iceland is reinforced at the same time that it is universalized. The specific collective knowledge of a people, its cultural capital or genetic commons, becomes the basic material for universal concepts of health. Culture and information become divorced from a specific place, experience, and time, becoming a global resource to be mined through "bioprospecting." The information society draws on the diversity and cohesion of societies globally for very specific information to create competitive asymmetries – while simultaneously establishing a universal framework of information technology in which specific information is valued and traded. The process, while using new tools, re-creates with information technologies and social information very old practices: prospecting, mining, surveying. Yet these networks are not fixed or finite like the previous models but dynamic, incorporating constant inputs of blood, records, and history that are directly derived from the lived experience of the Icelanders. Common events such as birth and death, sickness, and marriage become the basic material for competition and development in the global environment (see Figure 6.3).

Figure 6.3

Global patterns operating in Iceland

General patterns	Specific aspects	Specific form in Iceland
Informational processes, products, and industries	• Tacit, innovation-driven, and informational processes • Products embedded and defined by social knowledge • Global markets and regional, networked production for industries	• "Bioprospecting" for unique genetic patterns using skilled researchers, software, and hardware • Development of new drugs derived from Iceland's unique culture and regional industry characteristics • Global biotechnology industry linked with local firms and institutions to access and market local genetic information
Institutional mechanisms, norms, and flows structuring the informational environment	• Immigration • Investment • Innovation networks • National and regional policy • Private-public partnerships • Entrepreneurial firms–global firm linkages	• Skilled immigrant networks • Ties to US venture capital • National development initiatives • DeCode-government partnership • DeCode–global pharmaceutical firm links
Local processes and structures interacting with the informational environment	• Regional economic and social networks • Unique local knowledge/information • Regional governance institutions and capacity • Unique social capital • Unique cultural practices • Regional educational and scientific institutions	• Population homogeneity • Widespread and in-depth genealogy practice • Centralized state-controlled medical records covering most of the twentieth century • Tissue samples of the national population through state-controlled medical system • Returning immigrant researchers and doctors

As such, the information society is a merging of old and new, local and global, personal and social. This merging, as is clear in the case of Iceland, is universal while it is extremely specific and local. It takes material practices and makes them into immaterial, informational, digital forms of knowledge while seeking to draw on and supply these localized material practices. It does this by building on previous institutional forms and practices (the construction of centralized, public medical services) that interact with new forms (digitalized global biotechnology aimed at individualized medicine). Understanding the impact of this means understanding these processes and how they form into new institutions that structure social behaviour. In this way, the questions of privacy and access are really smaller questions derived from what institutions, laws, and norms will govern and structure the information society and economy.

Implications for Information Societies: Local Strengths and Global Trends

The case of Iceland opens up a window through which to consider both how IT and global trends are initiated and sustained over the long term. Both aspects create dichotomies that need to be carefully understood. The case of Iceland clearly highlights this tension. National legacies, both institutional and cultural, played a key role in producing the genetic resources that the nation drew on to establish a credible and viable information initiative, as the global environment evolved to value exactly these resources. The dichotomy for Iceland is that biotechnology, by its unique production and organizational forms, creates new social tensions and contradictions that challenge the very institutions and social norms that supported the creation of the genetic and social resources initially. Simultaneously, biotechnology, in its increasing importance as a central global industry, its information-based production process, its ability to absorb and valorize the specific local knowledge that Iceland possesses, also offers an excellent, in terms relative to other industries, chance for long-term economic and social development.

Clearly, a region's economic, political, cultural, and social inheritance shapes how a region will interact with the global environment. The structure of this interaction depends on the local adaptation of resources to address the challenges arising from the very interaction with broader economic and social trends. The interesting aspect of Iceland is that its very strength and social cohesion are what make it such an attractive location for global information industries. How this social cohesion is drawn upon, not only as a global commodity but also as a local balance to global trends, is a central issue.

The opportunity around information, and around industries following informational production patterns, is that local knowledge is needed and valorized within the production process itself. This means that nations or regions are really developing a mastery of information-based economic and social processes and not information as an industry per se. Building a successful environment to manage these new social structures means creating synergy between various sectors of the economy and society that can consciously generate the resources to manage and synthesize local needs with global structures. The opportunities of IT (as process, product, and industry) are that, while local knowledge is valorized on a global level, the flows of information and knowledge can go both ways: global skills can be applied to local needs, or local norms can become standard global practice. This is the unique opportunity provided by an information-focused strategy, if the global-local relationship can be structured to achieve this end. As such, even simple choices – an opt-in or opt-out system, the database's technical structure, the openness of the debate itself, the form of regulation and modification – can have wide-ranging impacts within the global environment.

Understanding Iceland's development within these new global structures is directly linked to evaluating how well Iceland understands these opportunities and challenges inherent in its genetic and biotechnology initiative. The case outlined above seems to question whether government officials and local entrepreneurs fully understand the dichotomies and opportunities of an information-focused environment that is simultaneously local and global. In other words, the very process of establishing the database has built institutional bridges between global trends and local social structures. The concern is that the policies in Iceland reflect a failure to understand that such linkages are being built. While it may be impossible to capture the benefits of the global environment without such linkages, not recognizing that such linkages exist means a limit on the ability of Iceland to structure such relationships around local and public, rather than private or global, priorities.

Understanding the impact of biotechnology is intimately linked to tracing how local and global interactions become institutionalized and structured within the information society. Overall, the Genetic Database strategy in Iceland is an ideal example of how interactions with the global environment transform both the local and the global simultaneously. Iceland presents a case where the personal links to the global, where local information, culture, and practice are transformed into a global resource, and where the impact of IT as process, product, and industry shapes new social structures. It is exactly such combinations that will increasingly come to matter in other regions and other IT industries. The institutions and norms that come to be established around early interactions, as in Iceland, foreshadow the possibilities and challenges of things to come.

Notes

1 Based on the $200 million contract with Hoffmann-LaRoche or the valuation of deCode at its initial public offering (IPO).

2 See Haraway (1991, Chapter 1) and Haraway (1997, Chapters 1 and 2) for further elaboration on this.

References

Agre, Phillip. 1995a. "Institutional Circuitry: Thinking about the Forms and Uses of Information." *Information Technology and Librarians* 14 (4): 225-30.

–. 1995b. "From High-Tech to Human Tech: Empowerment, Measurement, and Social Studies of Computing." *Computer Supported Cooperative Work* 3 (2): 167-95.

Agre, Phillip, and Douglas Schuler. 1997. *Reinventing Technology, Rediscovering Community: Critical Explorations of Computing as a Social Practice.* Palo Alto: Ablex.

Armour, Phillip G. 2000. "The Case for a New Business Model." *Communications of the ACM* 43 (8): 19-22.

Berlinski, David. 2000. *The Advent of the Algorithm: The Idea that Rules the World.* New York: Harcourt.

Billings, Paul R. 1999. "Iceland, Blood, and Science: Biotechnology and Its Social Implications." *American Scientist* 87 (3): 199-200.

Biotechnology Industry Organization. 2001. *Guide to Biotech.* On-line at <www.bio.org>, retrieved 27 June 2001.

Bowring, Finn. 2003. *Science, Seeds, and Cyborgs: Biotechnology and the Appropriation of Life.* New York: Verso.

Castells, Manuel. 1989. *The Informational City: Information Technology, Economic Restructuring, and the Urban-Regional Process.* Oxford: Blackwell Publishers.

–. 1996. *The Information Age: The Rise of the Network Society.* Cambridge, MA: Blackwell Publishers.

Central Bank of Iceland. 2003. *The Economy of Iceland.* Reykjavík. On-line at <http://www.sedlabanki.is>, retrieved 13 March 2004.

Clarke, Adele, and Virginia Olesen. 1999. "Revising, Diffracting, Acting." In Clarke and Olesen, eds., *Revisioning Women, Health, and Healing: Feminist, Cultural, and Technoscience Perspectives,* 3-48. New York: Routledge.

deCode Genetics Inc. 2004. On-line at <http://www.database.is/> or <www.decode.is>, retrieved 13 March 2004.

Dicken, Peter. 1998. *Global Shift: Transforming the World Economy.* London: Paul Chapman.

Eischen, Kyle. 2000. "Information Technology: History, Practice, and Implications for Development." Center for Global, International, and Regional Studies, University of California at Santa Cruz, Working Paper 2000-4. On-line at <http://cgirs.ucsc.edu/>, retrieved 13 March 2004.

–. 2003a. "Opening the 'Black Box' of Software: The Micro-Foundations of Informational Technologies, Practices, and Environments." *Information Communication and Society* 6 (1): 57-81.

–. 2003b. "Mapping the Micro-Foundations of Informational Development: Linking Software Processes, Products, and Industries to Global Trends." In S. Krishna and Shirin Madon, eds., *The Digital Challenge: Information Technology in the Development Context,* 297-317. London: Ashgate.

Fumento, Michael. 2003. *Bioevolution: How Biotechnology Is Changing Our World.* San Francisco: Encounter Books.

Geertz, Clifford. 2000 [1973]. *The Interpretations of Cultures: Selected Essays.* New York: Basic Books.

Gordon, Richard. 2001 [1994]. "State, Milieu, Network: Systems of Innovation in Silicon Valley." Center for Global, International, and Regional Studies, University of California at Santa Cruz, Working Paper 2001-3. On-line at <http://cgirs.ucsc.edu>, retrieved 13 March 2004.

Greely, Henry T. 2000. "Iceland's Plan for Genomics Research: Facts and Implications." *Juremetrics* 40: 153-91.

Gudmundsson, Adalsteinn. 2004. "Research on Aging in Iceland: Future Potentials." *Mechanisms of Ageing and Development* 125: 133-35.

Haraway, Donna J. 1991. *Simians, Cyborgs, and Women: The Reinvention of Nature.* New York: Routledge.

–. 1997. *Modest-Witness@Second-Millennium.FemaleMan-Meets-OncoMouse: Feminism and Technoscience.* New York: Routledge.

Harvey, David. 1990. *The Condition of Postmodernity.* Cambridge, MA: Blackwell Publishers.

Held, David, Anthony McGrew, David Goldblatt, and Jonathan Perraton. 1999. *Global Transformations: Politics, Economics, and Culture.* Stanford: Stanford University Press.

Henderson, Charles W. 1999. "Government Approves Plan to Sell Iceland's DNA." *World Disease Weekly Plus,* 1 February. On-line at <www.newsrx.com>, accessed 13 March 2004.

Kranzberg, Melvin. 1985. "The Information Age: Evolution or Revolution?" In Bruce R. Guile, ed., *Information Technologies and Social Transformation,* 35-53. Washington, DC: National Academy Press.

Kunzig, Robert. 1998. "Blood of the Vikings." *Discover* 19 (12): 90-99.

Lessig, Lawrence. 1999. *Code and Other Laws of Cyberspace.* New York: Basic Books.

Marshall, Eliot. 1998. "Iceland's Blond Ambition: Genetic Research on Icelanders." *Mother Jones* 22 (3): 53.

Masood, Ehsan. 2000. "Gene Warrior." *New Scientist* 167: 42-45.

Merz, Jon F., Glenn E. McGee, and Pamela Sankar. 2004. "'Iceland Inc.'?: On the Ethics of Commercial Population Genomics." *Social Science and Medicine* 58: 1201-9.

Mitchell, William J. 1998. *City of Bits: Space, Place, and the Infobahn.* Cambridge, MA: MIT Press.

Moukheiber, Zina. 1998. "Genes for Sale: Genetic Research in Iceland by deCode Genetics Inc." *Forbes* 162: 203-5.

Negroponte, Nicholas. 1995. *Being Digital.* New York: Alfred A. Knopf.

"Norse Code: Bioprospecting in Iceland." 1999. *Economist* 349: 99-100.

Rifkin, Jeremy. 1998. *The Biotech Century: Harnessing the Gene and Remaking the World.* New York: Jeremy P. Tarher-Putnam.

Saxenian, AnnaLee. 1994. *Regional Advantage: Culture and Competition in Silicon Valley and Route 128.* Cambridge, MA: Harvard University Press.

Senituli, Lopeti, and Margaret Boyes. 2002. "Whose DNA? Tonga and Iceland, Biotech, Ownership, and Consent." Paper presented at Australasian Bioethics Association Annual Conference, Adelaide, 14-16 February.

Shreeve, James. 1999. "Secrets of the Gene." *National Geographic* 196 (4): 42-75.

Statistics Iceland. 2004. *Iceland in Figures 2003-2004.* Vol. 9. On-line at <http://www.statice.is>, retrieved 13 March 2004.

UK Research Councils, E-Science Initiative. 2004. On-line at <http://www.researchcouncils.ac.uk/escience/>, retrieved 10 February 2004.

Weber, Max. 1968. *Economy and Society: An Outline of Interpretive Sociology.* New York: Bedminster Press.

Webster, Frank. 2002. *Theories of the Information Society.* 2nd ed. New York: Routledge.

Williams, Raymond. 1958. *Culture and Society, 1780-1950.* London: Chatto and Windus.

Winslow, Ron. 2004. "Gene Is Linked to Higher Risk of Heart Attacks and Strokes." *Wall Street Journal,* 9 February, B8.

Zucker, Lynne G., Michael Darby, and Marilynn Brewer. 1998. "Intellectual Human Capital and the Birth of U.S. Biotechnology Enterprises." *American Economic Review* 88 (1): 290-306.

7

Biotechnology and Social Control: The Canadian DNA Data Bank

Neil Gerlach

What comes to mind when we think about biotechnology? Most likely, we think about new reproductive technologies, gene therapy, genetically modified foods, the creation of hybrid animals, cloning, and the patenting of genetically modified life forms. These are the types of biotechnologies that have inspired intense public debate and media attention, primarily because of the unprecedented way in which they enable scientists to intervene in the processes of individual bodies. Less well known, but more widely employed within many Western societies, is another set of biotechnologies – DNA "fingerprinting" and DNA data banks. These intervene at the level of the social body to survey and record genetic information about populations of criminals. Many people have a vague awareness of these technologies, primarily through their use in a number of well-publicized cases such as the O.J. Simpson case in the United States and the David Milgaard and Guy Paul Morin cases in Canada. Fewer people are aware of the fact that the Canadian criminal justice system has actively employed these technologies since 1988 and has provisions for DNA warrants by which the police may seize bodily materials from suspects as well as a national DNA data bank for permanently storing biological samples and digitized genetic information on criminal offenders. These developments have already had significant impacts on criminal justice in particular and on social control in general. Despite their significance, however, there has been very little public discussion about the implications of DNA data banks and DNA testing.

What is the nature of the DNA data banking and warrant system in Canada, and how did it come to enter the justice system in its particular form? What are the enabling conditions that have allowed a potentially revolutionary technology of social control to enter so quickly and unassumingly into Canadian society? What are its implications for the present and future management of justice? In this chapter, I explore these questions, arguing that social conditions such as fear of crime, pervasive surveillance, redefinition of criminality, and rationalization of criminal justice institutions have opened the door for the relatively unproblematic entry of DNA technology into processes of social control. The result is a potential shift in criminal justice toward greater emphasis on social order and less emphasis on individual human rights.

Origins of DNA Testing

On two occasions, in 1983 and again in 1986, young women were sexually assaulted and murdered, each in neighbouring towns in Leicestershire, England. Something had changed in the intervening years, however – a revolution in crime detection. In 1985, Alec Jeffreys and a team of geneticists at the University of Leicester developed a technique for DNA testing of forensic evidence. All men in the three surrounding communities were asked to submit to DNA testing, but no matches could be found to the seminal material left at the crime scenes. The case was resolved when someone overheard one man telling others in a pub that he had been paid to stand in for another man in the DNA testing. Upon questioning, the police learned the identity of the man who had eluded the DNA net – Colin Pitchfork – who was apprehended, tested for a DNA match, and became the first person convicted through the use of DNA matching in the history of criminal justice.[1]

Basically, DNA typing compares the DNA profile obtained from tissue or bodily fluid samples found at a crime scene with that of a suspect to determine if they match at certain predetermined locations on the genome. Currently, the Canadian DNA data bank records information from thirteen loci within the genome. These loci are not known to provide any meaningful genetic information other than the identity of the person. There are currently three technologies of DNA testing available to a laboratory, including the Restriction Fragment Length Polymorphism (RFLP) method, the Polymerase Chain Reaction (PCR) amplification procedure, and Short Tandem Repeat (STR) analysis.

The RFLP method, invented by Alec Jeffreys, involves chemically extracting DNA from a sample and employing enzymes called restriction endonucleases to cut the DNA into fragments. These fragments are placed into a sieving gel and separated according to size by the application of an electric current. Once separated, the fragments are blotted onto a nylon membrane, combined with radioactive DNA probes that bind with complementary base pairs, and exposed to x-ray film to produce images that resemble supermarket bar codes. In this way, DNA fragments are rendered visible to scientists and can be compared to other DNA samples to see how well the "bar codes" match. PCR amplification adds replication technology to the process, allowing scientists to create billions of copies of a DNA strand within hours, drastically reducing the size of sample needed for testing. STR is a more advanced and more accurate method than RFLP analysis and is the most widely used today. It capitalizes on the fact that DNA contains repeating blocks called short tandem repeats, which occur in differing lengths and which vary in type and length in each individual. As well, the frequency of different STR lengths varies among different ethnic groups. Through the use of PCR to amplify lengths of DNA that contain STRs, a profile is created based upon the differences among individuals due to repetitions of some sequences.

Determining a DNA match is only the first step in the process of producing an identification of a criminal. The second involves the application of population genetics and statistics to the DNA sample. What is the probability that a match would occur with a randomly chosen member of the population? In other words, what are the odds that someone else within a given population might have the same pattern at the DNA loci tested in the laboratory? It is this calculation that provides much of the power of DNA evidence in the courtroom, with odds of one in several billion against a random match being quite common.

DNA Evidence in Canada

From the very beginning, it was clear that DNA testing had great potential for criminal identification and crime control. It was also clear that it would enter into criminal justice in interaction with a set of citizen rights that may be violated by the full use of the technology. The first question opened by the presence of this technology was how to obtain samples from suspects to match with crime scene samples. Do the police have the authority to seize samples without consent, or must they find other means? Between 1988, when the first Canadian case involving DNA testing was heard, and 1994, when the Supreme Court of Canada ruled on this issue, the state of the law was highly uncertain. Finally, in the case of *R. v. Borden* (1994), the Supreme Court ruled that there is no common law or statutory authority to permit the taking of bodily substances from suspects without their consent. The police would have to return to rummaging through garbage cans and ashtrays to obtain cells for DNA testing.

This situation was not acceptable to police or to government policy makers, and legislation was already under development at this time to clarify the situation. In June 1995, Parliament passed Bill C-104, An Act to Amend the Criminal Code and the Young Offenders Act (Forensic DNA Analysis). Bill C-104 established a procedure by which the police could apply to a provincial court judge for a DNA warrant to seize one of three types of bodily substances – hair roots, saliva and mouth swabs, and blood samples – while investigating certain designated offences. These are primarily serious offences of violence and sexual assault. Although the legislation is remarkable in Canadian law for the invasiveness of the procedures that it permits, it seeks to satisfy Charter of Rights requirements by setting out certain conditions for the issuance and execution of a DNA warrant. Section 487.05 (1)(a), (b), (c), and (d) of the Criminal Code instructs judges to issue DNA warrants only when there are reasonable grounds to believe that a designated offence has been committed, that a bodily substance has been found at the crime scene, that the offence was carried out by the person, and that forensic DNA analysis of a bodily substance from the person will provide evidence on whether the substances found at the crime scene came from the person. Bodily substances may only be obtained by trained personnel, and the police are required to advise the suspect that a DNA sample is being taken, by what authority, and that the DNA information can be used as evidence. There are additional considerations when the suspect is young.

Having addressed the first question of how to obtain samples from suspects, the next question involved determining the conditions under which these samples could be stored for future reference. When should they be stored? Whose samples can be stored? For how long should they be kept? Who should have access to them? What should be stored – biological samples, digitized information from the samples, or both? Beginning in early 1996, public consultation began, culminating in the 1998 passage of Bill C-3, the DNA Identification Act. In this act, the government authorized the establishment of a national DNA data bank to be maintained by the RCMP. It authorizes courts to order persons convicted of designated offences to provide samples of bodily substances for DNA analysis and storage within the "convicted offenders index" of the data bank. There is also a "crime scene index" to store samples and information found at crime scenes. Information from the two indices can be compared to find matches.

Bill C-3 also expanded the list of designated offences, which are now divided into primary and secondary offences. Primary designated offences include the most serious violent and sexual offences, where DNA evidence is most likely to be useful in solving the crimes. Secondary designated offences are serious but lesser offences for which the Crown must apply to the court for retention of a sample from a convicted offender. This leads to a substantial widening of the net in comparison to the number of designated offences in Bill C-104. Samples will be retained indefinitely, unless the offender is later exonerated or the conviction is overturned on appeal. In these circumstances, the samples will be destroyed, although the judge has discretion to retain the samples under certain circumstances. The legislation is also retroactive; currently sentenced offenders may also be sampled if they have been declared "dangerous offenders," if they have been convicted of more than one sexual offence, or if they have been convicted of more than one murder committed at different times. Access to DNA samples is strictly limited to personnel directly involved in the operation of the data bank. Samples may only be used for forensic DNA analysis, and only the name of the person from whom the sample is taken will be communicated to authorities involved in the investigation and prosecution of the offences. Criminal penalties apply to anyone who violates these conditions of privacy. In 1999, an addition was made to the DNA data bank provisions through Bill S-10, a Senate bill, which adds certain military offences to the list of designated offences and provides for a Senate review of the legislation after five years.

Finally, in July 2000, the last step was taken, and the National DNA Data Bank was officially launched in Ottawa. Overall, the provisions of the data bank and the warrant are products of a series of debates and fears expressed in the policy-making process and are carefully crafted to survive the inevitable Charter challenges they will face.

The policy discussions around the establishment of DNA warrants and the National DNA Data Bank are an interesting example of how a new biotechnology enters society. A number of groups participated in the public consultations that occurred around these laws and raised serious concerns about what they mean for state power over citizens' bodies. However, unlike many other types of biotechnologies, DNA testing and data banking have not become controversial in the public sphere. Some critics have argued that the normalization of DNA data banks within society could lead to increased levels of surveillance (Privacy Commissioner 1997), increased invasions of bodily privacy (Nelkin 1993), and the formation of a genetically based justice system (Friedland 1998). These potentials have not moved the public, in general, to become concerned about the issue. What are the social conditions that enable the entry of DNA testing and data banking into Canadian society in a relatively depoliticized manner?

I argue that the DNA provisions of the Canadian justice system have taken on their particular character due to the interaction of two forces: on the one hand is a set of human rights provisions and protections set out in the Charter of Rights and Freedoms and the criminal law; on the other is a set of social forces currently influencing public opinion about risk and security. These forces include a fear of crime, the emergence of a surveillance society, changing definitions of criminality, and processes of technical rationalization in crime management agencies. An examination of submissions made during public consultations by police organizations, victims' rights groups, civil rights groups, law associations, and representatives of the solicitor general's office demonstrate that these social forces form the axes around which much of the debate, controversy, and justification for the new technologies occurred. These are the enabling conditions that have allowed genetic technologies to enter Canadian criminal justice so easily.

Social Risk and the Development of DNA Banking

Fear of Crime

It is a truism among criminologists that, although crime rates declined in Canada during the 1990s, fear of crime increased. According to the 1996 International Crime Victimization Survey, the percentage of Canadians who feel fairly or very safe walking alone in their area at night is 73 percent, the third lowest rate of the eleven participating countries (Canadian Centre for Justice Statistics 1999). The international average among participating nations was 77 percent. The same survey was conducted in 1992, and Canada was one of five countries participating in both surveys. Among the other four – England and Wales, Finland, the Netherlands, and Sweden – there was no significant change. In Canada, the rate dropped by 5 percent (Canadian Centre for Justice Statistics 1999).

Although most Canadian citizens feel safe in their neighbourhoods, a significant and growing proportion do not. This is despite a 3 percent drop in victimization rates in Canada in the same 1992-96 period (Canadian Centre for Justice Statistics 1999). That rate remained stable up to the end of the 1990s (Besserer and Trainor 1999). That drop has not translated into a public perception of declining crime rates. Criminologists disagree over why this is happening, with much attention focused on media representations of crime and the effects that they may be having on public perceptions of security and risk. The classic research is that of Gerbner and his associates (1980), who argued that over time the themes and content of mass media, especially television, "cultivate" a common social reality for the audience. One aspect of this "reality" is a "mean worldview" produced through exposure to televised violence and characterized by a fear of crime out of proportion to actual crime rates. Gerbner's research has been criticized for its methodology and its findings (Graber 1980; Sacco 1982; Surette 1984); however, research in this "media effects" tradition continues, examining media representations of offenders, victims, and crime statistics and their impacts on public perceptions (Chiricos, Eschholz, and Gertz 1997; Ericson, Baranek, and Chan 1989).

A more profound critique of the media effects research on fear of crime is paradigmatic rather than methodological. There is a growing body of work attempting to understand the multidimensionality of fear of crime and to avoid labelling it as simply an irrational emotional response to media representations. Although media representations definitely play a role in creating a hyperreality of crime, fear of crime stems from other social factors as well: community disorganization and breakdown of sociability (Walklate 1998), increasing anxiety and a sense of insecurity in a period of rapid governmental, economic, and technological change (Hollway and Jefferson

1997; Sparks 1992), and a breakdown of relationships of trust among citizens and between citizens and criminal justice experts (Garland 1996; Lupton and Tulloch 1999). In other words, fear of crime is a manifestation of anxieties created by the shift to postmodernism and the concurrent destabilization of social institutions. Interaction between media representations and changing social experience creates in the individual a feeling of unease and insecurity that translates into an enhanced fear of crime.

Given this pervasive distrust of the public sphere, fear of crime is a primary enabling condition for the full embrace of genetic technologies in criminal justice, technologies that promise faster and more precise identification of criminals. This is one of the implications of a national survey conducted in 1995 by COMPAS, an Ottawa-based research firm, which found that, of the 1,004 respondents, 88 percent agreed that suspects, and not just convicted criminals, should be forced to give DNA samples to the data bank (Thanh Ha 1995). This suggests that Canadian citizens are willing to empower law enforcement agencies to enhance surveillance over the population, even if it means a general decrease in the genetic privacy of citizens.

Further evidence of the impact of fear of crime is the occurrence of at least two "DNA sweeps" in Canada during the 1990s – investigations in which the RCMP asked all men in a community to submit to DNA testing in order to remove themselves from the suspect list. The first occurred in Vermilion, Alberta, where three women had been sexually assaulted within four years. The second occurred in Port Alberni, British Columbia, where a young girl had been sexually assaulted and murdered. Testing was entirely voluntary, but an incident in Vermilion suggested that refusal to submit to testing may not be an easy option. In a town hall meeting convened to discuss the RCMP request, two men protested the DNA sweep on the ground of privacy, and they were angrily shouted down by the other townspeople. Informal processes of social control intervened to compel compliance with the "voluntary" testing. Incidents such as these imply a willingness to surrender individual rights for the sake of public security within social conditions defined as risky. The potential problem with DNA sweeps is that they may invert the presumption of innocence. In the absence of relations of trust among citizens, people become afraid of the public sphere and of each other, increasingly looking to state authorities to impose social order.

Surveillance Society

A second reason why DNA testing and data banking do not elicit strong public questioning may be the fact that the technology does not really intervene at the level of the individual body, nor is it about transforming bodies. Rather, it intervenes at the level of the social body through surveillance and coding. As such, it enters into society through preexisting relationships between citizens and surveillance technologies. This is not to say that Canadians are unconcerned about, or unaware of, surveillance technologies. Both the press and the entertainment media have made citizens highly aware of real and potential privacy violations through means such as monitoring of cellular telephone calls and e-mail, consumer databases created through recording credit card transactions, and video camera surveillance of public and private places. However, in matters of criminal justice, fear of crime appears to discourage public action to restrict state and private use of surveillance technologies such as DNA data banks, opening the door to the formation of a surveillance society.

What is a "surveillance society"? In recent years, discussion of this concept has focused on the notion of the "panopticon." This term was originally coined to describe an architectural design featuring a central observation tower surrounded by cells arranged in a circular fashion to allow for complete observation by a watcher in the tower. Ideally, the watcher would be invisible to the occupants of the cells, with the result that they would never know if they were being watched and would always have to assume that they were under observation. The panopticon was originally designed for use in workplaces and prisons, but the French theorist Michel Foucault (1979) appropriated the notion as a metaphor to describe the dominant form of social control within the modern world. He argued that modern institutions are organized in space and time to render populations and their processes observable – a surveillance society. In the twentieth century, electronic technologies allowed for an enhancement of this process (Mehta and Darier 1998).

Gary Marx (1988), in his analysis of undercover police work, outlines the contemporary characteristics of surveillance. Today surveillance is invisible, involuntary, capital rather than labour intensive, decentralized into self-policing, oriented toward observing whole categories of persons and not just individuals, and more intensive and extensive than in the past. The result is a shift away from the direct use of force to maintain social control in favour of manipulation rather than coercion, computer chips rather than prison bars, and remote and invisible tethers rather than handcuffs and straitjackets. In other words, panopticism in the public sphere does not proceed through a prison-like atmosphere of surveillance as in George Orwell's *Nineteen Eighty-Four*. Rather, it operates through the embedded nature of surveillance technologies. Our everyday transactions are silently and unobtrusively entered into databases. As a population, we have long been accustomed to being affixed with numbers such as social insurance numbers, employee numbers, and credit card numbers. We are no longer the unruly mobs of nineteenth-century cities; rather, we are a normalized, fashion-conscious, educated, and well-behaved populace that exists within a "superpanopticon" of pervasive surveillance (Poster 1990, 126). The public has been disciplined into participating in its own surveillance, and in that context the DNA data bank is simply another kind of database in an information society where such technologies are administratively normalized and accepted. From the perspective of the Canadian public, the possibility of enhanced genetic surveillance within the general population seems to be a relatively small price to pay in exchange for the risk-management capacities of the DNA data bank.

Even within the short life of the DNA warrant and data banking provisions, there have been significant expansions in the ability of the state to maintain surveillance over the Canadian population. For example, with the DNA warrant provisions, the government has granted itself unprecedented powers. Under common law, an attempt by the state to seize bodily substances from a citizen would be a violation of "bodily integrity" and would constitute an assault, except in extreme circumstances. The DNA warrant provisions have the effect of replacing the doctrine of bodily integrity and authorizing as much force as necessary to execute DNA warrants.[2] In addition, Bills C-3 and S-10 each marked an expansion of the designated offence list originally established in the DNA warrant provisions. Offences for which a convicted person may be entered into the data bank range from serious acts of violence to relatively minor and common offences such as assault, robbery, breaking and entering, and any attempts to commit these offences. In 1995, about 85,000 individuals were charged with common assault and about 48,000 charged with break and enter. Thus, many relatively minor offenders could have their DNA added to the data bank, which could eventually encompass a large segment of the Canadian population (Privacy Commissioner 1998). If these developments are any indication, there appears to be a tendency by authorities to expand the surveillance capabilities of the DNA data bank by empowering police to take samples and by periodically lengthening the list of designated offences.

Redefining Criminality

Within modernity, questions about the fundamental causes of crime have often revolved around the nature/nurture debate. Among criminologists, psychologists, and sociologists, the dominant view over the past two centuries defines criminality as a product of one's living conditions, socioeconomic life chances, and social group's moral climate. This has also been the perspective held by the law, which assumes that at some point, even if someone has lived in poor social conditions, criminal action is a question of free will; it is a matter of individual morality or self-government, which may be restored through therapeutic rehabilitation. Increasingly, however, social scientists are questioning their earlier assumptions as current research shows that socioeconomic status is not as significant a factor in criminality as once was thought (National Council of Welfare 2000). At the same time, research consistently indicates that a very small number of criminals are responsible for most violent crimes (Gibbs 1995).

Findings such as these have reignited interest in the "nature" side of the debate, and increasingly government policy makers, social scientists, and biologists are redefining criminality in terms of biological inclinations. This approach is not new, and there have been periods in history when biological explanations for criminality have been given considerable credence, only to be ultimately rejected on the basis of research and because of the political implications of their assumptions. At the turn of the past century, for example, eugenic theories of criminality – belief that criminality and violence are hereditary traits – became very important in social policy, resulting in as many as fifteen states in the United States passing laws allowing for the involuntary sterilization of convicted criminals (Kevles 1997). Proponents of eugenic theories of criminality included notable researchers such as Charles Davenport and Cesare Lombroso, both of whom based their theories on the phenotypes of individuals, arguing that certain individuals were "atavistic" throwbacks to more primitive people and could be identified by visible features such as longer arms, heavy jaws, and narrow foreheads. From this perspective, such people are born criminals and cannot be rehabilitated.

We appear to be entering another period of biological determinism in criminal justice, particularly in the United States, where biological disorder is increasingly offered as a defence in criminal cases. Ranging from the influence of sugar, premenstrual syndrome, postpartum depression, and the possession of an extra Y chromosome, these defences are used to argue that defendants acted involuntarily and, therefore, do not deserve punishment (Nelkin and Lindee 1995). Legal analysts, such as Steven Friedland (1998), suggest that the question now is not whether genetic evidence will be admitted into court but when and under what circumstances.

What would be the outlines of a genetically based legal system? We can speculate on this question based upon certain current trends. An immediate effect would be a shift in notions of responsibility. If we define certain forms of criminality in terms of genetic imperatives, can these criminals be held responsible for their actions? Arguably not, and if that is the case one of the basic assumptions of our criminal justice system – culpability based upon free will, autonomy, and individual responsibility – must be reconsidered. Guilt or innocence in certain cases would no longer be based upon normative considerations but upon physiological ones, and, furthermore, genetic predisposition would be a defence to future dangerousness. These are issues already faced by courts in cases where genetic defences have been argued.

A second implication of a genetically based justice system would be a shift in emphasis from identifying criminals who have committed criminal acts to identifying deviants who have the *potential* to commit crimes. If links can be found between genetic factors and violent or antisocial behavioural tendencies, experts may have the tools needed to predetect those who are inclined toward certain crimes and place them under observation and control to ensure that their genetic tendencies do not manifest in a harmful manner. Predetection of criminality has been a quest of modern criminology for some time and periodically leads to undesired consequences. For example, in the 1960s, Scottish researchers believed they had found a link between an extra Y chromosome in males and a higher level of aggressiveness that could manifest as violent crime. Eventually, research in Europe and North America showed no link between an extra Y chromosome and crimes of violence, and the matter died down. In the meantime, a number of men who had never committed crimes had been stigmatized as potential criminals (Hubbard and Wald 1993).

A third implication might be a virtual abandonment of the rehabilitation ideal in criminal justice – the idea that offenders can be rehabilitated in the criminal justice system and return as contributing members of society. This notion is already under heavy attack, and even current government policy is oriented toward rejecting the possibility of rehabilitation and toughening sentences. Rejection of the rehabilitation ideal has resulted in the exploration of a number of new options in criminal justice or a return to older options. For example, the three strikes rule in many states in the United States is premised upon the idea that some people are born criminals and cannot be rehabilitated; therefore, after their third chance, they should be permanently incarcerated. Amitai Etzioni (1999) argues that, when sex offenders are released from prison, they should be placed in guarded communities where they would live out their lives without the opportunity to reoffend. Some opponents of the death penalty have expressed concern that DNA testing will inevitably move criminal justice toward acceptance of the death penalty since DNA matches appear to strongly indicate guilt or innocence and may eventually reveal genetic inclinations toward violent crime (Easterbrook 2000). One of the arguments against capital punishment has always been uncertainty of guilt in many cases and the moral wrong of putting an innocent person to death. It is assumed that because of the precision involved in DNA matching that much of the uncertainty associated with identifying individuals as present at the scene of the crime is reduced.

DNA data banks enter into this process as a potential initiating technology. Fears about the role that the Canadian DNA data bank may play in the development of a genetic justice system were voiced by the federal privacy commissioner in a presentation to the Standing Committee on Justice and Human Rights in 1998: "Retaining a databank of genetic samples from convicted offenders will inevitably attract researchers who want to analyze the samples for purposes that have nothing to do with forensic identification. This scientific curiosity, coupled with growing pressure to reduce crime by whatever means, no matter how intrusive, will almost certainly lead to calls to use samples to look for genetic traits common to 'criminals.' This type of research, while perhaps of scientific interest and possible social value, raises complex legal, ethical and moral problems that we have yet to resolve" (4). A number of concerns are expressed in these comments. There is a concern that fear of crime is leading to reduced attention to individual privacy. There is a concern over the absence of moral constraints on scientific research, and finally there is recognition that genetic research is increasingly informating the human body and the human person, codifying and illuminating the fundamental building blocks behind our physical and psychological selves. The DNA data bank can play a significant role in these processes, and the argument is that we must ensure that appropriate limits are placed on its use. Will we place these limits on the DNA data bank, or is it the first step toward a genetic justice system?

Technical Rationalization

A fourth enabling condition for the entry of genetic technologies into criminal justice is the ongoing process of technical rationalization. In the early twenty-first century, this takes the form of corporatizing many of our social institutions (McBride and Shields 1997; Ritzer 1996; Shields and Evans 1998). Today the corporation has become the exemplary institution upon which others are modelled. This "corporatizing" involves a number of elements. First, organizations must restructure along corporate lines to become "flexible" in the face of increasingly chaotic market processes in the global economy. Second, employees must transform their identities to become entrepreneurs – people who act upon individual initiative and resources in order to add value to the organization. Third, organizations must become thoroughly informated, utilizing information technology not only to increase the efficiency of internal operations and communications but also to open direct channels to "customers" in order to determine their desires and adapt to them. Fourth, an ethic of competition should pervade the organization as a motivating force toward ever greater profit and success. This is the basic model of organization that applies not only to corporations but also to schools, universities, hospitals, government departments, and even volunteer organizations.

Increasingly, the corporate model also applies to criminal justice institutions. They are under the same social pressures to be efficient, cost effective, and competitive while redefining members of the public as "clients" (Heward 1994). These pressures manifest in a number of ways. The nature of police work is changing as police increasingly become information managers in a policing system that utilizes community resources to survey community problems, but also in the sense of providing risk information to insurance companies, regulatory agencies, financial institutions, health and welfare institutions, and motor vehicle agencies (Ericson and Haggerty 1997). Community policing is, in part, about shifting the role of the police to that of information management in order to serve better consumers/citizens. Increasingly, police managers utilize a market-based language of shifting the police *force* to a police *service,* adopting mission statements, managing organizational culture, and implementing reward systems for excellent "customer care" (Heward 1994).

In terms of the court system, Nils Christie (1996) notes the importation of business rationales into the functioning of the courts. He argues that the central values are becoming the clarification of goals, production control, cost reduction, and division of labour, all combined with the coordination of these functions at a higher level of management. We can see this in courtrooms, which increasingly take on the character of a business office. Lawyers' gowns are gone, the language becomes more straightforward, seating arrangements are more similar to those of administrative decision makers, the computer becomes a natural fixture. Solemn, ritualized ceremonies of justice are exchanged for convenience and efficiency. We also see this drive to rationalize the justice system in its outputs. Production is faster in that more people can be sentenced with less effort than before. Decisions are more uniform, and punishments are more "equal." Bottlenecks have been removed, plea bargaining ensures fast confessions, and sentencing tables, which automatically prescribe sentences for certain crimes, ensure fast and predictable decisions on punishment.

DNA data banks and warrants seem to be ready-made for a rationalized, technologized justice system. Certainly in government policy statements about the need for a DNA data bank, "keeping pace with technological advances" is a recurring phrase. Other efficiency-based reasons are given as well, including focusing investigations, shortening trials, increasing the number of guilty pleas, deterring offenders from committing serious offences, and keeping up with other countries, especially the United States (Solicitor General 1996; Department of Justice 1995). Here we see a very direct importation of an industrial logic into the framing of the data bank as the government draws upon a language of savings, managerial efficiency, technological imperatives, globalization of crime detection, and, ultimately, an ethos of crime management.

The potential problem with a strong emphasis on rationalizing criminal justice is that it ignores the essential ritualism that lies at the heart of social justice. This was an observation made by Émile Durkheim (1893) over a century ago when he argued that crime serves a positive function within the social organism; it provides opportunities for social members to engage in rituals of reflection and punishment that reinforce the norms and morals essential to social solidarity or work to reform them. If crime management occurs within laboratories, police computers, and efficient, office-like courtrooms, the public is further removed from the process, and the rituals that assert moral imperatives are eliminated. Testimony, witnessing, argumentation, and the exercise of discretion while taking into account larger social and moral questions tend to be lost as laboratory technicians enter the courtroom and simply state whether or not there is a DNA match. Richard Ericson and Kevin Haggerty (1997) point out that in such a system "Deviance is thus merely a contingency for which there are risk technologies to help spread the loss and prevent recurrence. It becomes a technical problem which requires an administrative solution, rather than an occasion for expressing collective sentiments and moral solidarity" (40).

The criminal justice system is the institution in which the state exercises force over citizens in the most direct manner. It is imperative that citizens feel some degree of responsibility for, and involvement in, this system to ensure that state power is restrained by public sentiment. Does the rationalization of the system, in part through DNA technology, operate to further remove citizens from criminal justice processes in the name of efficiency and productivity?

Potential Impacts of DNA Data Banks

Although the National DNA Data Bank has been in full operation only since July 2000, DNA identification technologies have been employed in the Canadian criminal justice system for well over a decade. There are indications of the types of impacts that their presence may have on the long-term shape of crime management as they interact with the social conditions outlined above. DNA data banks and tests intervene in a preexisting relationship between citizens and the state and offer potentials for altering certain aspects of that relationship. The nature of this alteration is determined in part by the social forces that currently influence public attitudes toward crime control and social control in general – fear of crime, pervasive surveillance, redefinitions of criminality, and the rationalization of criminal justice along a corporate model.

When implementing criminal justice policy, democratic governments must always address a fundamental axis of governance: social control versus individual rights. Where policies lie on this spectrum is a strong indication of the nature of the relationship between citizens and the state. The federal government in Canada has attempted to balance social control and individual rights by limiting the data bank to convicted offenders and by requiring that police obtain warrants before taking DNA samples by force. However, certain aspects of the DNA data bank and warrant laws and certain incidents that have occurred around this technology suggest that it may have already opened the door to greater social control and less emphasis on individual rights. Overturning the doctrine of bodily integrity, expanding the list of designated offences, and conducting DNA sweeps should give rise to some public concern about the potentials of DNA technology and the temptations faced by criminal justice personnel to expand the use of the technology. Certainly, the Canadian government and the police have powers and panoptic resources that they have never had before.

In the meantime, fear of crime, a comfort with surveillance, a redefinition of criminality, and the rationalization of crime control institutions combine to depoliticize this expansion of state power made possible by DNA data banks. At the very least, these conditions ensure that there is little protest over the implementation and expansion of a significant surveillance technology as society looks to a technological solution for a social problem. DNA testing does produce impressive results compared to other forms of evidence. Although DNA testing is not infallible, its accuracy rate is very high, resulting in the prosecution of a number of dangerous offenders and the release of wrongly convicted people. This enhancement of accuracy has been bought, however, at the price of increased state power over the citizen's body. It will be essential for critical social scientists to monitor this new technology over the next few years to analyze how it continues to contribute to changes in state power and criminal justice into the future.

Notes

1 For an interesting and detailed account of the Colin Pitchfork case, see Joseph Wambaugh's *The Blooding,* a novelization of the investigation and the first use of DNA matching technology in a criminal investigation.

2 Authorization to use as much force as necessary in carrying out a DNA warrant is set out in section 487.07(1)(e) of the Criminal Code. This is not technically the first provision allowing state agents to enter a citizen's body. In the previous year, Parliament added sections 254(3) and 256 to the Criminal Code, allowing officers to seize blood samples in cases of suspected impaired driving. The case law on these provisions tended to set very strict limits on their application.

References

Besserer, S., and C. Trainor. 1999. *Criminal Victimization in Canada*. Ottawa: Statistics Canada.

Canadian Centre for Justice Statistics. 1999. *The Juristat Reader: A Statistical Overview of the Canadian Justice System.* Toronto: Thompson Educational Publishing.

Chiricos, T., S. Eschholz, and M. Gertz. 1997. "Crime, News, and Fear of Crime: Toward an Identification of Audience Effects." *Social Problems* 44 (3): 342-57.

Christie, N. 1996. *Crime Control as Industry: Towards Gulags, Western Style*. London: Routledge.

Department of Justice, Canada. 1995. *Obtaining and Banking DNA Forensic Evidence*. Ottawa: Department of Justice.

Durkheim, É. 1964 [1893]. *The Division of Labor in Society*. New York: Free Press.

Easterbrook, G. 2000. "DNA and the End of Innocence." *New Republic* 223 (5): 20-21.

Ericson, R., P. Baranek, and J. Chan. 1989. *Negotiating Control: A Study of News Sources*. Toronto: University of Toronto Press.

Ericson, R., and K. Haggerty. 1997. *Policing the Risk Society*. Toronto: University of Toronto Press.

Etzioni, A. 1999. *The Limits of Privacy*. New York: Basic Books.

Foucault, M. 1979. *Discipline and Punish: The Birth of the Prison*. New York: Vintage.

Friedland, S. 1998. "The Criminal Law Implications of the Human Genome Project: Reimagining a Genetically Oriented Criminal Justice System." *Kentucky Law Journal* 86: 303-66.

Garland, D. 1996. "The Limits of the Sovereign State: Strategies of Crime Control in Contemporary Society." *British Journal of Criminology* 36 (4): 445-71.

Gerbner, G., L. Gross, M. Morgan, and N. Signorielli. 1980. "The Mainstreaming of America: Violence Profile No. 11." *Journal of Communication* 30: 10-29.

Gibbs, W.W. 1995. "Seeking the Criminal Element." *Scientific American* March 1995: 100-7.

Graber, D. 1980. *Crime News and the Public.* New York: Praeger.

Heward, T. 1994. "Retailing the Police: Corporate Identity and the Met." In R. Keat, N. Whiteley, and N. Abercrombie, eds., *The Authority of the Consumer*, 240-52. London: Routledge.

Hollway, W., and T. Jefferson. 1997. "The Risk Society in an Age of Anxiety: Situating Fear of Crime." *British Journal of Sociology* 48 (2): 255-65.

Hubbard, R., and E. Wald. 1993. *Exploding the Gene Myth.* Boston: Beacon Press.

Kevles, D. 1997. *In the Name of Eugenics: Genetics and the Uses of Human Heredity.* Cambridge, MA: Harvard University Press.

Lupton, D., and J. Tulloch. 1999. "Theorizing Fear of Crime: Beyond the Rational/Irrational Opposition." *British Journal of Sociology* 50 (3): 507-23.

McBride, S., and J. Shields. 1997. *Dismantling a Nation: The Transition to Corporate Rule in Canada.* 2nd ed. Halifax: Fernwood Publishing.

Marx, G. 1988. *Undercover: Police Surveillance in America.* Berkeley: University of California Press.

Mehta, M.D., and E. Darier. 1998. "Virtual Control and Disciplining on the Internet: Electronic Governmentality and the Global Superpanopticon." *Information Society* 14: 107-16.

National Council of Welfare. 2000. *Justice and the Poor.* Ottawa: National Council of Welfare.

Nelkin, D. 1993. "The Social Power of Genetic Information." In D. Kevles and L. Hood, eds., *The Code of Codes: Scientific and Social Issues in the Human Genome Project*, 177-90. Cambridge, MA: Harvard University Press.

Nelkin, D., and M.S. Lindee. 1995. *The DNA Mystique: The Gene as a Cultural Icon.* New York: W.H. Freeman.

Poster, M. 1990. "Foucault and Data Bases." *Discourse* 12 (2): 110-27.

Privacy Commissioner of Canada. 1997. *Annual Report.* Ottawa: Canada Communications Group.

–. 1998. *Remarks to the Standing Committee on Justice and Human Rights Concerning Bill C-3, the DNA Identification Act.* Ottawa: Privacy Commission.

R. v. Borden, [1994] 3 S.C.R. 145, 92 C.C.C. (3d) 404 (S.C.C.).

Ritzer, G. 1996. *The McDonaldization of Society.* Rev. ed. Thousand Oaks, CA: Pine Forge Press.

Sacco, V. 1982. "The Effects of Mass Media on Perceptions of Crime." *Pacific Sociological Review* 25: 475-93.

Shields, J., and B.M. Evans. 1998. *Shrinking the State: Globalization and the Public Administration of "Reform."* Halifax: Fernwood Publishing.

Solicitor General of Canada. 1996. *Establishing a National DNA Data Bank: Consultation Document.* Ottawa: Solicitor General.

Sparks, R. 1992. *Television and the Drama of Crime: Moral Tales and the Place of Crime in Public Life.* Buckingham: Open University Press.

Surette, R., ed. 1984. *Justice and the Media: Issues and Research.* Springfield, IL: Charles C. Thomas.

Thanh Ha, T. 1995. "DNA-Testing Law Passed." *Globe and Mail,* 23 June, A1, A7.

Walklate, S. 1998. "Excavating the Fear of Crime: Fear, Anxiety, or Trust?" *Theoretical Criminology* 2 (4): 403-18.

Wambaugh, J. 1989. *The Blooding.* New York: Bantam.

8
Biotechnology as Modern Museums of Civilization

Annette Burfoot and Jennifer Poudrier

> The critical history of collecting is concerned with what from the material world specific groups and individuals choose to preserve, value and exchange.
>
> – James Clifford, 1998

The History of Collecting

Collecting is both a powerful and a fluid social dynamic. Although it is not usual to see the current biotechnological revolution as a form of collection, it is. Genes from all types of life forms are now gathered for analysis and manipulation. Probably the best-known contemporary biotechnology collection is the human genome "library." There are also collections of seeds and other plant genetic material held by commercial interests such as the multinational corporation Monsanto. Genetic engineering techniques help to fuel these modern but private museums of natural history as a simple change to a life form's genetic structure enables a patent to be held on the resultant "mutant." In contrast, some people have started countercollections to try to preserve original genetic stock from genetic engineering and the corporate control of living entities. An examination of the biotechnology revolution as an extension of collecting activities that extends back to preindustrial Europe sheds light on the rapidly growing field of contemporary bioengineering and provides a critical ability to respond.

To understand what collection means, we need to explore its history, especially in regard to modern social practices. These practices are tied to both the European Renaissance in art and the scientific revolution and are implicated in Europe's economic and political expansion. Although we usually focus on cultural artifacts or "art" in terms of collection, we will see how early museums were devoted to both natural science collections and collected art. Museums have a particular history that links to ancient concepts of the classification of human activity and the understanding of what being human means. Although we know little of ancient Greek and Roman collecting, we do know of their ideas about how the world was parcelled in terms of relative worth. This classification process is an essential component of the act of collecting. You need to know what merits finding and holding on to before you can collect it. It also serves as the basis for modern collections of art and science.

The modern museum exhibits both art and science collections. Consider the Montreal Museum of Fine Arts, the Canadian Museum of Civilization in Hull, and Alberta's Tyrell dinosaur museum. This practice reflects the ancient Greek classification of the arts that included medicine, drama, agriculture, and war. What were not included in the ancient concept of liberal arts were the "servile arts" performed by a lower class of people (often producing goods and services enjoyed by the upper classes). For example, the upper classes would write and attend plays but would rely on a lower class of actors to perform them. During the European Renaissance, which coincided with the scientific revolution, a new form of classification occurred. The social status, morality, and intellectuality that characterized the liberal arts in ancient societies were replaced by a system that emphasized gratification over intellect. Paintings and sculptures emphasized forms and imagery that pleased the aristocracy controlling the production of arts. During the Renaissance, European sculptures, paintings, and tapestries filled first private and then semi-public collections and galleries. For example, the highly prized marble statues of Gianlorenzo Bernini were collected by the Borghese pope Paul V at the Villa Borghese just outside central Rome. The pope entrusted his nephew, Cardinal Scipione Caffarelli Borghese, with diplomatic and ceremonial relations for the papal court that were to take place at the villa amid this incredible collection of exquisite artwork. In this early museum, the skill of Italian artists and the cultural superiority of the collector were displayed simultaneously (Fiore 1997, 5).

The effects of European expansion, colonization, and increased global trade were also apparent in the growing art collections throughout Europe. Surveys of art throughout most of the twentieth century usually centre on European art as definitive and tend to include other art as only bookends of a sort. Otherness is first realized temporally as prehistoric art: that which is the farthest from the European tradition of culture in terms of time. The main exception to this rule is classical Greek and Roman culture that is also removed in time. But these cultural histories are usually included as the precursors of European art. The Renaissance (which literally means rebirth) is heavily indebted to these ancient cultural traditions and forms. Otherness as geography comes at the end of the artistic survey as so-called primitive art. Here art forms are gathered by location (African, Asian, etc.) and often by function (ceremonial masks, Native garb, etc.) but rarely, if ever, as the work of acclaimed individuals or as ideals of human expression and identity, as with European artists and art (Gardner 1975). These prehistoric and primitive artworks were included in the modern museum collection, again, as spectacle and as a possession of the dominant culture. One of the more famous examples of cultural appropriation through collection is found in Pablo Picasso's painting *Les Demoiselles d'Avignon* (1906-7). The painting's striking facial images are a result of Picasso's visit to an exhibition of a collection of African masks in Paris. Picasso incorporated these foreign faces in a way that has drawn both acclaim and criticism (for debasing both femininity and Africanness) (Duncan 1998).

Simultaneously with the Renaissance, science rose as a new and independent social undertaking with classifications of its own. Scientific taxonomy, or the practice of classifying natural phenomena, also has roots in ancient Greek and Roman culture. Aristotle was one of the first to generate a classification system based on a mixture of philosophical and empirical approaches to understanding the natural world. He ordered what he observed into categories defined by similarities (genus) and distinguished by difference (differentia). His system relied on determining what a living thing was "by nature," which meant a deep-seated aspect of being that could not be changed in contrast to superficial aspects. For example, all birds have wings and reproduce by laying eggs, but their plumage varies greatly. This system of logic served as biological or natural classification until the 1800s. In 1758, Carolus Linnaeus, a Swedish botanist, used the Aristotelian philosophy of ordered nature to develop a method (binomial nomenclature) to order and name the living natural world (plants and animals) ("Taxonomy" 1985).

With this paradigm of classification in hand, and with the loose application of genus and species to all of the arts, Europe began to establish collections of both the natural and the cultural worlds as common social practice. Naturalists such as Ulisse Aldrovandi (1522-1605) and Athanasius Kircher (1602-80) are described as typical members of a newly established profession of scientists who approached nature "as a collectible entity," a possession, and their museums were where they could "bring all of nature into one space" (Findlen 1996, 1). Natural history museums sprang up in the seventeenth century all over Europe, particularly in the cradle of scientific debate at this time: Italy. In the university city of Bologna, you can still visit many of these early exhibition sites, which have remained largely unchanged over hundreds of years. For example, you can find Cesare Taruffi's (1821-1902) collection of anatomical "monsters" preserved in alcohol and formalin or as wax models. The monstrous referred to the abnormal and would include two-headed sheep, babies with severe spina bifida born without skulls, and so on. The purpose of these displays was primarily educational, and they served as a sort of visual library for doctors and scientists. The Museum of Human Anatomy established in 1742 by Pope Benedict XIV contains wax models of women's reproductive organs, including very pregnant uteri, to train physicians and midwives.

These museums became proud sites of items collected not only from the local natural world but also from exotic faraway places. As communication and transportation expanded rapidly in this period of European economic expansion and colonization, examples of foreign flora and fauna (including humans) began to fill the museums. These collections of scientific artifacts from exotic places became conflated with collections of cultural artifacts from the same places. The science of anthropology, developed during this period, exemplified the profound connection between cultural and scientific collection. Anthropology was devoted to the collection of other social and cultural artifacts for study and display "back home." Thus, the display of a stuffed pygmy monkey from the deep, dark jungles of Amazonian South America was effectively the same as that of mysterious ceremonial masks from East Africa. Difference, both locally as monstrous occurrence in normal affairs and as foreign exotica, became more than a component of classification; it became the source of fascination, a form of public entertainment, and a way to make a profit (Fusco 1998).

The modern systems of artistic and scientific classification remain with us today, albeit with some significant changes. We tend to view the respective collecting activities as separate even though they continue to share fundamental principles, such as logical hierarchies. Also, scientific classification has recently shifted. The Linnaean system based on the initial work of Aristotle forms the basis of contemporary biological classification, but modifications have occurred since. Charles Darwin's work in evolutionary theory in the mid-1800s and discoveries of microscopic life such as viruses in the 1900s have altered how life is ordered. More recently, the role of evolution (the survival of the fittest as an explanation of how the living natural world has evolved) has come under criticism as a key component in biological classification (Denton 1996). Also, the emphasis in genetic research and engineering, especially the recent completion of mapping an entire human genome, points to new ways of classifying life as microscopic code rather than Aristotle's vague notion of essential being. And in a return to art and science's shared history in terms of classification, one recent approach even likens genetic development to painting (Coen 1999).

Despite these innovations, the basis of modern collection found in seventeenth-century European museums can be found in contemporary science practices, especially with the advent of a new genetic terrain that presents as largely an unexplored and potentially rich territory. We will examine four key aspects of biotechnology in terms of the principles of modern collection: the human genome project, the control of seeds, biological preservation, and medicine (including pharmaceuticals). But first, in order to understand how contemporary collection works, we need to understand more about the dynamic of collection itself.

Criticisms of Historical Collecting

James Clifford (1998) provides a historical critique of collection practices, with a particular focus on the way in which tribal artifacts and cultural customs are transformed through Western processes of collection, exchange, classification, and display. On their own, collections are neither self-sufficient nor innocent. They are always fluctuating and transgressing through categories of beauty and authenticity: categories defined by the collector. They are also highly influenced by the values of the collector and transformed through ideological priorities involved in processes of collecting.

Clifford describes the way in which Western ideological, political, and institutional collecting traditions appropriate, relocate, redefine, and "exotify" non-Western objects. For example, objects are first appropriated from their source and relocated to museums or exhibitions, under the purview of (supposed) objective and scientific collection practices. They are then decontextualized from their source in terms of time, coherence, and completeness and recontextualized, classified, ordered, and assigned value according to Western ideological terms. Likewise, the redefinition of objects also transforms the source from which they came whereby non-Western cultures and peoples become redefined and essentialized as "exotic" and "tribal" through the same ideological classification practices as were applied to the artifacts themselves.

Douglas West (1994) shares similar criticism of the way in which contemporary collections do not reflect the worldview, continuity, or wholeness of the context from which the artifacts came. Paying particular attention to the relationship between Aboriginal peoples of Canada and the museums that display and describe Indigenous artifacts and cultural traditions, West criticizes museums because objects "remain disconnected from living things" and become redefined and exotified in the context of museums (363). For example, he discusses the contemporary use of phrases such as "wilderness" and "exotic" to describe Indigenous lands. He states that, "from Indigenous world views, there are no wildernesses at all, just homelands, and Nature is never considered exotic. Wilderness and exotic are Euro-Canadian terms" (364).

Clifford (1998, 94) contends that objects and cultural artifacts are both collected and given value through "disciplinary archives and discursive traditions." These priorities, like the Enlightenment values of progress, production, and scientism, are concerned with rationality, efficiency, linear modes of description, immutable definitions, and exchange value. As such, through the ideological transformation of objects, Western cultures, in fact, collect themselves. In other words, collections of exotic objects reveal more about the priorities of the collector than they do about the objects themselves, and they represent the cultural traditions of the West rather than those of the culture and context from which they came. What we gather around ourselves defines us; alternatively, what defines us is what we gather around ourselves. As West (1994, 365) eloquently puts it, "each culture has its own way of keeping the collective culture ... We must remember that the medium of that collection is, as McLuhan argued convincingly, the message. In other words, museums are another element of the scientific world view itself."

In order to expose and clarify the ideological nature of Western collections, Clifford (1998) recommends that the "art-culture system" should not be taken for granted but opened up and described. Moreover, processes of collecting ought to be visually displayed as part of the exhibitions themselves. In this way, exhibition spectators may become more aware of the way in which Western traditions appropriate and order the world. As Westerners, we might also become more self-critical about our roles in the appropriation, classification, exchange, display, and control of other cultures.

Principles of Collection

Drawing on the history of collecting practices, we can delineate certain principles and compare them to contemporary biological collection and classification. This comparison will help to exemplify the extent to which genetic science is not pure and objective but a reflection of the ideological and political nature of society. Genetic collections and the way in which they are represented are social constructions. Principles of collection common to both historical collecting and contemporary biotechnology include the ideal of social progress as a motivating factor, assembling the collection through isolation and then appropriation, categorizing and presenting the collection according to dominant knowledge systems, and assigning economic and exchange value to collectables.

Social Progress

The purpose of collecting generally has to do with social progress and improvement according to Enlightenment traditions of scientism, empowerment, and emancipation. The development of knowledge follows specific formulations of social progress that have become an essential component and a defining feature of Western societies. For example, in modern systems of exchange, social progress is often measured by the extent to which societies are able to define, identify, make valuable, rationalize, or put things under the control of scientific and economic systems.

As it relates to social progress through scientism, the "art-culture system" and genetic science share conceptual ideals. Both are epistemologically purposive, rational, reductionist, and efficient. While art-culture systems produce an efficient description of artifacts and cultural practices according to Western theories of cultural formation, genetic science provides an efficient means to codify life and life processes as strands of protein that are comprehensibly described and categorized according to the logic of empiricism. Both systems of knowledge formation rely on the assumption that such ordering improves control of large and complex collections. This conceptual control represents social progress through scientific method.

Moreover, both art collection and DNA collection focus on a transformation from the primitive and exotic to the progressive and rational. For example, exotic artifacts and foreign cultural practices have often been described in derogatory terms such as "savage," "barbaric," "turbulent," "impure," "vulgar," "unintelligible," "incomprehensible," "fascinating," "enigmatic," "mysterious," and "immaculate." Similar descriptors are used to refer to incomprehensible genetic information, including the enigmatic "book of life." Likewise, genetic scientists are often referred to as "gene hunters" who aim to track down and conquer the primitive genetic code or as cartographers mapping a mysterious genetic wilderness.

Appropriation

Collection necessarily entails removing objects or knowledge from their original context to the one where they are rearranged and redefined according to dominant ideologies: appropriation. "Appropriate" is a Latin derivative of the word *property,* which literally means "to make one's own" (Clifford 1998). Besides dislocating artifacts, collecting dehistoricizes or rehistoricizes the knowledge and cultural systems from which the artifacts and specimens originate. The local conceptualization of the object's power, its mystery, its meaning are stripped from it when it is removed from its context and given a new meaning according to foreign standards. Also, appropriation essentializes collected materials and knowledge by determining and generalizing which characteristics signify objects as exotic artifacts.

Classification and Display

Once artifacts are in hand, classification begins and includes the marking of domains and the reordering of things into discrete and linear categories. The ordering of the collection and the manner of its display are discursive and "speak" a story as told by the collector and not the collected. "A scheme of classification is elaborated for storing or displaying the object so that the reality of the collection itself, its coherent order, overrides specific histories of the object's production and appropriation" (Clifford 1998, 97). For example, artifacts are displayed in a specific proximity to other objects deemed similar so that the characteristics of one object are redescribed in the context of the other. This classification system is often arbitrary and lacks the context and depth of the cultures or contexts from which they all came.

Drawing on Susan Stewart's work (1985), Clifford (1998, 97) also identifies how "collections – most notably museums – create the illusion of adequate representation of a culture by first cutting objects out of specific contexts (whether cultural, historical, or intersubjective) and making them stand for abstract wholes." Metaphors are used in display as explanation or cultural links from the dominant culture to the appropriated one. This leaves only the vision of the collector's worldview and not an objective or transparent view of the appropriated culture. Display also provides an opportunity to profit from a collection as a form of entertainment.

Value

Economic, cultural, and scientific value is created through processes of defining and ordering. Clifford (1998, 96) suggests that cultural identity is made up of the things that are collected and that such collections are organized according to powerful but "arbitrary systems of value and meaning." Since systems of value are not stable and are ideological, the value of collections is not objective and pure. Value is also related to "authenticity," whereby, in the making of meaning, value and conceptions of authenticity are interdependent. On the one hand, the value of an object is dependent on its authenticity, purity, or naturalness; on the other hand, the characteristics that define its authenticity are dependent on standards of value prioritized by presupposed collection systems. For example, the exchange value, the cultural significance, or the scientific merit of an artifact is produced by characteristics of authenticity, rareness, or uniqueness that are conversely determined by given systems of worth. Powerful distinctions are made at different points of history, within a certain ideological frame. This frame or general system of categorization constitutes the types of objects that are collected, the standards by which they are organized, and the ways in which they are represented or displayed.

Like the collection of art, the collection of genetic information is similarly related to value – both cultural and economic. Economic value is evident in the issue of ownership and control of discovered and collected genetic information. Values usually associated with art or cultural collecting – authenticity, rareness, and the taming (classification) of the exotic (savage) – also apply. This process, the collection and redefinition of genetic information, like the collection of art, is inherently ideological.

Contemporary Biotechnology Collections

The term "biotechnology" has come to represent a complex array of interests and applications. This array includes a rapidly expanding area of research in science and technology, a burgeoning industry, and a multitude of practical applications. The science journal *Nature Biotechnology* devoted its entire first issue in the new millennium to the question of who controls biotechnology (2000). Increases in mutual funds worldwide in the year 2000 were led by gains by biotechnology companies ("Dire Time" 2000). And public debate was dominated by concerns over genetically modified food, privacy, and genetic discrimination. Biotechnology has drawn perhaps more public reaction than any other development directly involved with science and technology. This reaction is neither narrowly focused nor uninformed and involves a sophisticated public appreciation of the potential risks involved with, for example, the genetic modification of food. There is also a clear public understanding of the role of politics and multinational corporations in the biotechnology sector.

Biotechnology companies led all categories in mutual fund earnings in June 2000, with a return of 12.8 percent; the top performer was C.I. Global Biotechnology Sector Shares (returning 28.3 percent) ("Dire Time" 2000). Another company of note, especially in regard to public reaction, is Monsanto. The firm has drawn considerable public attention in its heavy-handed attempts to promote and protect its interest in genetically modified plants. Early in 2000, Monsanto signed a publicly contested contract with the University of Manitoba worth $7 million to establish a research site on a campus in Canada's agricultural heartland (Paskey 2000). This mirrors similar contracts between biotechnology companies and universities in the United States. The Michigan Economic Development Corporation has established a $1 billion, twenty-year program to develop "major hubs of high technology industry" on public university campuses, where high-tech refers to "bioengineering and other emerging life-science industries" (Schmidt 2000). Biotechnology is a rapidly changing area and requires large amounts of research and development with scientists and engineers. This industrial requirement coincides, especially in Canada, with massive and chronic public underfunding of universities. Thus, increasingly, Canadian universities' research agendas are driven by commercial interests, with biotechnology as the current commercial leader.

There is also much concern over the politics embedded in biotechnology. Now that Canada and the United States recognize the patenting of plant (and some animal) life, genetic engineering takes on the contentious role of creating mechanisms of direct corporate control over food types not only as a product but also as potential. Genetic anthropologists are now collecting valuable plants to intentionally mutate the organisms for increased yield and profit. This genetic appropriation is also being used to "protect" nearly extinct or threatened species, plant and animal, including human types. An international consortium of twenty-eight interested countries is seeking to establish the Global Biodiversity Information Facility to enhance centres of biological information and collections concentrated in developed countries, including natural museums, with a system of interconnected databases. In contrast to information about biodiversity, the actual biologically diverse organisms extend throughout the world and are concentrated in developing countries. Plants have always served a medicinal purpose, and the collection of genetic plant materials serves a burgeoning pharmaceutical industry.

The promise of individually targeted medical treatment fuels gene therapy research. The television show *Biography* (2000) chose Craig Venter and Francis Collins, the two men credited with overseeing the completion of the mapping of an entire human genome, for the biography of the year 2000. This human genome project has reconstituted the human body into a form ready for genetic collection and manipulation. Health care is currently being reordered according to genetic classification. The development of genetic probes for gene-based illness and disease increases concerns, especially in the United States, over access to health care: if you are determined to become ill with a certain disabling disease, who will provide you with medical insurance to cover the cost of your care? Who will hire you? Will you be pressured not to reproduce (Gostin 1994)?

Plant Biotechnology

One of the earliest examples of plant biotechnology is the development in the early 1980s of the Ti plasmid in the bacterium *Agrobacterium tumefaciens* as a vehicle to carry foreign genes into the genetic structure of tobacco (Hodgson 2000, 29). Since then plant biotechnology has grown into a highly contentious and widespread phenomenon. It involves the genetic modification of food as well as the control of agriculture worldwide. Vandana Shiva (1997), a Third World agricultural activist, describes the shift as ideological and one that favours monoculturalism and seeks to undermine agricultural diversity, especially in developing countries. In general, there has been significant public resistance to plant biotechnology as well as promises by its proponents to reduce the use of pesticides and even eradicate world hunger. The main principles of collection underlying plant biotechnology are the decontextualization and appropriation of genetic variation and the designation of value through patent protection.

In seventeenth-century European natural museums, plant specimens were significant parts of the collections. Gherado Cibo of Italy collected and illustrated thousands of plants from the Marches area near his home in Northern Italy. His illustrations, the early modern way to represent nature, are praised in terms of how well he "captures" it (Findlen 1996, 166-67). Later, in the eighteenth and nineteenth centuries, plant collection became highly significant especially as the North West pushed its explorations and conquests farther into foreign territories where the majority of the Earth's biodiversity resides. At this stage, plants were literally captured and removed from their native lands for the pleasure and intellectual satisfaction of European explorers. Many of our common houseplants in the North West, such as the fern, reflect these early plant safaris. The plants and observations of plants gathered by these early naturalists served chiefly to build a growing library of plant biology housed in major urban centres of developed countries. These exotic creatures were also enjoyed by the social elite: orangeries (hothouses attached to manor houses to grow citrus fruit) became highly fashionable in eighteenth-century England. Throughout this period, plant life was being collected and coded according to the logical principles developed by Aristotle and refined by the new scientists of the modern era and delighted in by the European aristocracy.

With the advent of twenty-first-century biotechnology, this process of natural collection gains a sharper edge as the specimens are manipulated as well as decontextualized from their natural and cultural surroundings. One of the main effects of plant biotechnology is the rendering of nature in a much more propriety form than during early modern science. The motivation of early modern science was to broaden a scope of knowledge as a form of social progress. If commercial exploitation could be garnered along the way, then all the better, but it was not obviously the prime motivation for exploration. Scientists prided themselves as arbiters of truth and did not perceive themselves as swayed by commercial or any other concerns. Today it is difficult to distance science as any pure pursuit of knowledge from explicit commercial interest.

Northern and Western appetites, meanwhile, have become accustomed to foods not normally grown in local environments. Contemporary "gene hunters" are now trolling the genetic terrain and the actual globe for examples of genetic diversity in plants that may be useful to a variety of commercial applications but chiefly food production for the North West. Unlike earlier explorations for examples of biodiversity that sought to complete knowledge about the natural world, these explorations are looking for immediate and usually commercial application in the richer North West. Genetic manipulation of plant material involves the splicing of desirable genetic qualities into a plant. For example, Illini Xtra-Sweet has been modified to produce more sugar and to slow its conversion to starch so that it is more conducive to shipping (Pollan 1994, 50). Other food staples such as Roundup Ready soybeans have been genetically altered so that they will respond only to the herbicide manufactured by the company that also holds the plant patent (Suzuki 1999-2000). Even the methods of modification themselves are patented, allowing for an exponential reach of corporate control over plant life collections: the patented "terminator gene" controls gene expression or reproduction so that any seed generated from a plant with this gene cannot reproduce. Farmers can no longer collect seeds from previous crops and have to buy new seed each year ("US Patents" 1998). Thus, the interest in the collection of genetic variation for commercial exploitation dominates local interests in collecting seeds for self-sufficiency.

The entire genome for rice is about to be completed. One of the largest companies devoted to the financial exploitation of plant biotechnology, Monsanto, started the lengthy project of unravelling the rice genome and then published its findings to allow for others, including publicly funded researchers, to help complete the project. Although the release of this information was seen by some as contrary to normal patterns of commercial exploitation, Monsanto retained the right to share in any patented discoveries from the genome and, simultaneously, benefits from public research dollars ("Map" 2000). The purpose of mapping such a genome is to prepare for and facilitate the genetic manipulation of all rice plants. It is possible to splice foreign known genes with desirable traits into the plant, but the process is more successful when using genes from the same genome. Rice is one of the world's food staples; those who control the main strains through bioengineering and patenting will effectively hold millions of people's lives in their hands.

There are two interconnected issues with this increased monopolization of plant gene "banks." One is the ethics surrounding capitalization of plant life. The other is the effect of the monocultural ideology that drives plant biotechnology. The publicized motive for continued plant appropriation is to solve world hunger and make food production more effective and less dependent on pesticides. The purpose of collecting as many genetically diverse samples as possible, according to this logic, is to find genetic qualities, such as pest and drought resistance, to create new genetic combinations that will produce food where before it was difficult or impossible. The promise also includes greater yields and less dependence on pesticides. However, there is no talk of discontinuing the patenting of plant life, and companies will always argue that they need patent protection to cover the costs of their research and development (R&D). The estimated cost of biotechnology R&D for each of the past three years is $6 billion (Hodgson 2000, 31). It is understandable within the logic of capitalism that these costs should be covered through sales. However, within this logic is a contradiction in relation to the promise of plant biotechnology as a solution to world hunger: how can the enormous costs of R&D of the technologies be met if they are to be supplied without any or much cost to those who need it (see Broerse and Bunders, this volume, Chapter 3)? Also, experts in development assistance from Oxfam Canada (2000), Greenpeace, and CUSO claim that "biotech will not feed the world." They argue, along with many others, that the issue of world hunger is not one of supply but of delivery and cost – both matters determined by the North West.

The other issue is that of monocultural ideology. Shiva (1997) describes a complex interaction of human culture and diverse natural life in the management of sustenance in so-called undeveloped countries. She provides an illuminating comparison between local and dominant knowledge systems' respective attitudes to trees and food plants. Within local knowledge systems, there is a highly interconnected cycle between forest management (forests provide wood, food, water, fodder, and fertilizer) and agriculture (which relies on the water, fodder, and fertilizer to produce pulses, cereals, and oilseeds). Conversely, dominant systems that rely on plant monocultural biotechnology see forests as only a source of wood and agriculture for the exclusive production of oilseed, wheat, and rice crops (Shiva 1997). Local self-sufficient and environmentally sound practices are drowned out by global markets' demands for the food preferences of those who can afford them.

The public is very aware of the problems associated with the contemporary collection of plant life and the manipulation of its genetic components. Besides recontextualizing these components within an appreciation of native environments and cultures, this opposition evokes collection as subversion. "Botanical arks" and "blooming fruiting archives of genetic and cultural information" are terms used by those resisting the global control of plant production through seed monopolization (Pollan 1994, 50). These are attempts by individuals and small groups such as organic farmers and farmers from the South to protect genetic diversity from eradication and capitalization (de la Perrière, Ali, and Seuret 2000). The International Food Policy Research Institute has produced an Internet resource with information and analyses in over 2,000 documents on collective action and property rights (CAPRi). The site provides a search engine to access information on natural resource management, property rights, and activism (UNEP/GEF Biodiversity Planning Support Program 2000). Also, part of the broadly based resistance to genetically modified food is grounded in the criticism of the commodification of plant life. "Food security" is a term recently coined to identify the risks associated with the private and corporate collection and manipulation of plants and plant genetic information (Madeley 2000).

There are also forms of collecting that aim to conserve biological diversity. Biological diversity, or biodiversity, is defined by the World Resources Institute (2000) as the total interdependent variation of biological species, genes, and ecosystems on Earth that form the foundation of all life. In the past several decades, increasing levels of human consumption alongside a significant degree of commercial development and exploitation have seen the erosion of the Earth's capacity to support biological life. Biodiversity is jeopardized in a context where plant and animal species are becoming extinct at ever-increasing rates and existing plant and animal species are becoming more uniform and thus less adaptable (World Resources Institute 2000).

The strain on biodiversity is exacerbated in a current global socioeconomic context in which the world's poor majority struggle to survive and the affluent minority continue to consume at increasing rates. Moreover, transnational corporations continue to overexploit the biologically rich but economically poor areas of the world. The degree of commercial and other benefits from these resources has not been matched by a corresponding investment in sustainable development and the conservation of biodiversity (World Resources Institute 2000).

In recent years, there have been substantial efforts aimed at protecting the resources in these areas from exploitation and depletion (International Plant Genetic Resources Institute 2000). While conservation and protection obviously depend on extensive political, economic, and social deliberation, many international organizations have supported the collection of genetic material from a variety of traditional plants and animals. In particular, conservation in developing areas has been focused on preserving "landraces" or indigenous plant and animal species developed by traditional agriculturalists without the influence of modern technological intervention. For example, in the Philippines, local farmers' organizations sustain and exchange landrace root crops such as sweet potato, taro, and yam (Manicad 1996), and in Ecuador the International Centre for Agricultural Research in the Dry Areas (ICARDA) has supported the local development of landrace-based barley (International Plant Genetic Resources Institute 2000). These initiatives are expected not only to conserve local biodiversity but also to support economic development. Similar projects for the conservation of landrace germplasm exist in Canada. In Alberta, agronomist Sharon Rempel coordinates a group of individuals who share responsibility for conserving the biological diversity of landrace varieties of wheat. Monsanto has increased some of the strain that these individuals face by introducing, on a test plot basis, an experimental variety of Roundup Ready wheat. The task of preserving this germplasm is made more difficult due to the constraints that the Canadian Food Inspection Agency has on revealing the location of these test plots. Biotechnology companies are reluctant to reveal the locations of test plots due to concerns about protecting confidential business information and concerns that acts of ecoterrorism (e.g., pulling up crops) may occur.

Exotic and Near-Extinct Plant and Animal Preservation

There are many extinct and endangered plants and animals. According to the National Wildlife Federation, each year nearly 40,000 different species become extinct, and that rate is accelerating ("Facts" 2000). It is well known that, for the most part, humans are responsible for the dangers that many species of plants and animals face. The most typical factors that threaten different species include the loss of habitat due to the many devastating effects of human development and the commercial exploitation of natural environments. There is also considerable appropriation of wildlife through gathering, hunting, and fishing for commercial products or for sport (Wildlife Conservation Society 2000a). Biotechnology apparently provides a technological fix for extinction through the systematic collection and preservation of samples from threatened species. Aspects of genetic engineering also promise the fantastic possibility of reviving already extinct plant and animal types.

The beluga sturgeon of the Caspian Sea, once an abundant source of caviar, has recently been classified as endangered because of overfishing, loss of habitat, and human pollution (Wildlife Conservation Society 2000b). In

California, there are over 289 threatened or endangered species, and in Florida the Florida cougar (also known as the Florida panther or puma) is currently classified as a critically endangered species ("Endangered Species" 2000). The cougar is endangered because it is presumed to be a threat to livestock and is treated like common vermin by farmers. Moreover, it has been hunted for sport, and, because of massive human intrusion into its natural habitat, it is increasingly forced into dangerous areas (National Wildlife Federation 2000).

International and national efforts to conserve endangered species began to emerge in the 1960s during the various environmental or green movements. Contemporary organizations, including the World Wildlife Fund, the US National Wildlife Federation, and the influential World Conservation Monitoring Center, were established to collaborate and to develop policy aimed at the protection of threatened species. A large part of the task was to distribute information and to make all peoples aware of endangered species. Concurrently, a widespread populist concern for the environment, including endangered species, has emerged. For example, Bruce Cockburn, a Canadian activist and musical artist who wrote and performed the song "If a Tree Falls" (released in 1988), has dedicated a significant amount of time and talent to preserving the natural environment from corporate appropriation of natural habitats. Hollywood has produced films associated with the concerns about endangered species and the environment. For example, *Gorillas in the Mist* (released in 1988) tells the story of primatologist Dian Fossey's crusade to save a community of endangered gorillas from poachers and indifferent government officials. The 1993 film *Jurassic Park* also addresses the issue of extinction but with a twist. It presents a scenario of DNA "magic" whereby the genetic information found in a preserved blood sample was used to revitalize the dinosaur species. The blood was extracted from the gut of a mosquito that had been preserved for millennia in petrified tree resin called amber.

Fiction seems to mirror fact as scientists now pursue the possibility of cloning extinct animals from DNA samples. They attempted this with a preserved Siberian woolly mammoth but were unsuccessful. DNA used for cloning must come from live tissue samples, and, in the case of the woolly mammoth, the cells were not intact and the DNA was too fragmented to be useful for cloning (Gene Letter 2000). However, there have been many attempts to clone living but endangered species primarily as a means of preservation. These efforts are based upon the techniques used to clone Dolly the sheep by scientists at Scotland's Roslin Institute in 1997. The cloning procedure used in preservation attempts on a near-extinct Asian ox involved removing the genetic material from a regular cow's embryo and replacing it with that of the ox. An electrical current was used to fuse the cells and create an embryo, which was then implanted into a domestic cow. Other preservation attempts through cloning have been tried with a rare goat, the Indian cheetah, and the Tasmanian tiger. Scientists in Iowa cloned the Asian gaur ox ("How Noah" 2001). Gestated in and born from a surrogate cow mother at TransOva Genetics in Sioux Center, Noah died two days later from common dysentery ("Cloned Ox Dies of Natural Causes" 2001). This procedure is expected to generate a modern "ark" of endangered species.

In a region of northern Spain, the last-known survivor of a rare goat species was found dead, having been crushed by a tree ("Clone Plan" 2000). Fearing its extinction and with some hope in cloning technologies, veterinarians in the area extracted a tissue sample from the animal's ear. While the sample tissue containing the DNA has been preserved, environmental authorities have not yet decided if the goat will be cloned. Meanwhile, in China, there is hope to preempt the extinction of the panda, but cloning experiments are not going well ("Clone Hope" 1999).

Researchers in India are planning to clone the Indian cheetah with the expectation of reintroducing the animal into the wild. While the cheetah does exist in the wild in other territories, it was driven out of India some years ago ("India" 2000). Also, in Australia, researchers are addressing the possibility of cloning the extinct Tasmanian tiger (actually a carnivorous marsupial). The last-known living tiger died in captivity in 1936, but museum researchers have recently uncovered a sixth-month-old pup preserved in alcohol. Scientists may attempt to use the genetic information of the pup for cloning, possibly using a Tasmanian devil (also a marsupial) as a surrogate mother ("Tasmanian" 1999).

The cross-species implantation procedure worked for Dolly the sheep and will most likely work for future Noahs (both of which originated from the degeneticized egg of a domestic cow). However, cloning and producing live births are not simple. For example, in the case of Noah, scientists made 690 previous unsuccessful attempts. For other types of endangered species, there will be much more difficulty. Consider, for example, the panda, which is the only member of its genus with no obvious candidates for transplantation. The dilemma will be finding a suitable surrogate species that would not reject foreign genetic material ("How Noah" 2001).

Despite the fact that the development of cloning techniques provides some hope for preserving species on the brink of extinction, the approach is not proceeding without public debate. On the one hand, there is the obvious redemptive possibility of preserving endangered species. Cloning could provide a "genetic fix" for the devastating effects of the human encroachment on and exploitation of natural ecosystems. Even more, imagine the excitement and the knowledge generated if extinct animals like the woolly mammoth, the saber-toothed tiger, or dinosaurs were cloned and given new life. On the other hand, there are many other concerns. From the point of view of conservation specialists, there are good reasons to be wary of dependency on cloning techniques for the preservation of species. The World Wildlife Fund (2000) argues that, since the success of genetic innovation is very limited, cloning should never be considered a substitute for good conservation strategies. Even if certain species are given new life through cloning, there is no way to ensure the survival of whole populations. There is no guarantee that a healthy gene pool could be produced and sustained. Since species are most often threatened by loss of or damage to habitats, there is no assurance that natural habitats could be restored and maintained. For the World Wildlife Fund (2000), cloning does not address causes of endangering species and cannot provide the fix. Moreover, there is some concern regarding funding whereby supporting expensive cloning techniques diverts much-needed money from long-term, sustainable, and more holistic conservation strategies.

Several other pervasive issues emerge as they relate to the concepts of collection. Will cloning evoke the same kind of profit-driven interest in collecting the DNA of endangered or extinct plants and animals? Given the exotic and rare nature of certain genetic information (e.g., that of the extinct Tasmanian tiger), the scientific value of this collection is certainly considered priceless. On a more general level, will the rare genetic information of cloned animals see the same commercialization and monopolization of life that already exists in the case of plant biotechnology? Who will own the genetic information, and who will own the rights to use developed cloning techniques? In January 2000, Britain issued a patent for the cloning process that created Dolly (Gene Letter 2000). The patent was granted to Geron Corporation (a biopharmaceutical company that focuses on the development and commercialization of therapeutic and diagnostic medical products). However, what this means for the development of future cloning techniques and for the ownership of rare genetic material is not clear.

The issue of commercialization of species is further exacerbated by a new approach to classification. In December 1999, a German-based nonprofit organization, Biopat, was created to generate funds for the preservation of species whereby individuals can buy and name species that either have not yet been named or have been recently discovered. The species available range from orchids to toads to butterflies, and Biopat organizers suggest that one might immortalize a loved one by giving the gift of a species: "Can you think of a more unusual gift or a more individual form of dedication with which to honor, say, a member of your family or a personal friend?" (Biopat 2001). Biopat contends that the proceeds generated go toward developing taxonomical descriptions of biodiversity, systematically recording it, and ensuring its preservation. Criticism of the organization is based once again on the commodification of life whereby species are named for and by individuals with wealth rather than by the Linnean tradition of ordered and systematic classification and nomenclature. Some scientists call Biopat "critter colonialism" as a strategy to snatch up the animal world, others are worried about the contamination of the taxonomic discipline by accepting frivolous names, and others are concerned about the commercialization of living things ("You" 2000).

The issue of cloning is also related to the concept of reordering and redefining. In the case of playing with the genetic material from animals, redefining comes into sharper focus. Just recently, scientists in Portland reformulated and produced a baby monkey with a new gene that can cause cells to become fluorescent ("Monkey" 2001). Researchers are not necessarily interested in a glowing monkey (fluorescence has already been accomplished with other animals, including a rabbit); rather, they are interested in creating genetically modified monkeys that could reproduce as new transgenic colonies. With fluorescence as a test case, these monkeys would eventually be altered genetically to develop targeted human diseases. The monkeys would then be used, as is the patented "oncomouse," as a specially designed research subject to develop treatments and cures in humans. Despite the present scientific limitations of this research, the colonies of monkeys would be redefined at the genetic level. Their value might also become rearticulated into the public consciousness as a new type of test species created for the purpose of scientific research. How will we think of and treat these monkeys or other newly created species? Will they become defined by their purposive value? Under conditions where life is increasingly becoming commercialized and commodified, will newly generated species carry a certain price tag?

The case of the transgenic monkeys, like other efforts geared toward genetic engineering, has raised concern about its implications for humans. Can scientists clone human beings? After the successful birth of Dolly, researchers were very enthusiastic about the possibility of cloning humans. At the same time, there seemed to be a benevolent reason for cloning human beings, such as assisting infertile couples in conceiving children ("Race" 1998). In November 1998, South Korean researchers at the infertility clinic Kyunghee University Hospital in Seoul announced that they had cloned a human embryo. Claiming to have fertilized the egg of a thirty-year-old woman with a cell from elsewhere in the body, Dr. Lee Bo-yon stated that the embryo had divided into four cells before it was subsequently destroyed ("Scientists" 1998). Since researchers were concerned about the social and legal contingencies of the experiment, the trial was discontinued before the embryonic cells had divided any further. There was a great deal of skepticism from the international scientific community, which doubted the success of the experiment and wondered if the announcement did not contain some elements of publicity seeking. This announcement also stimulated much debate regarding the ethical issues of cloning humans, and since then there has been an international moratorium on cloning human embryonic cells. For example, the United Nations Educational, Scientific, and Cultural Organization (UNESCO) added and condemned human cloning in its 1997 "Declaration on the Human Genome and Human Rights." However, Dorothy Wertz, a leading scholar in social, ethical, and legal issues associated with genetic technology, explains that, despite present agreements not to clone human beings, there is no guarantee that future research will not include cloning (Gene Letter 2000).

The issues surrounding cloning humans come into sharper focus alongside ongoing scientific efforts geared toward understanding the function of genes and what constitutes the entire human genome. The Human Genome Project has become the primary hub of research in human genetics. It has also become the fundamental site of collection for human DNA and genetic information.

The Human Genome Project

The Human Genome Project (HGP) provides a fascinating example of the principles of collection, including social progress as it relates to scientism, appropriation, relocation, reordering, and classification. These principles will be used to analyze the HGP, but first the development of genetic science and the Human Genome Project itself must be described.

Early attempts to scientifically determine the function of human genes came at the turn of the nineteenth century when the scientist Sir Francis Galton (1909) proposed that human heredity could account for feeble mindedness, which led to crime, alcoholism, prostitution, and ultimately social degeneration. The optimistic union of genetic knowledge and social disorder, referred to as eugenic science, came to represent a form of social engineering. It entailed the progressive application of scientific knowledge to social malady and was expected to control social decay by encouraging the reproduction of the genetically superior and reducing the reproduction of the biologically inferior (Kevles 1985). At its core, eugenic engineering was a functional and instrumental effort geared toward social progress. It was intended to improve the human race through the generation and application of knowledge to the human condition by aligning human behaviour with scientific categories of normalcy/abnormalcy and genetically fit/unfit. By the 1940s, however, eugenic science as a discipline came to an end, not only because of its affiliation with horrific Nazi sterilization, birth control, and euthanasia laws, but also because the basis of the relationship between genetic composition and social degeneration was not scientifically proven (McLaren 1990).

With the 1953 discovery of the double helical structure of human DNA by James Watson and Francis Crick, the world of genetic science and its relationship to the human condition was revived and now proceeds as a lucrative step toward the betterment of humankind. The most substantial contemporary genetic research effort is the Human Genome Project. It has been referred to as the "search for the holy grail of science" (Lee 1991, 19), and its findings have been described as "the source book for biomedical science in the 21st century" (Department of Energy and the National Institutes of Health 1991, vii). The knowledge gleaned from the project is expected to "revolutionize biology and medicine [and] forever change how we understand the human body" (HGP 2000). In essence, the HGP is heralded as the quintessential key to modern medical, scientific, and social progress.

The HGP was officially launched in the United States in 1990 as a joint initiative between the US Department of Energy and US National Institutes of Health. While many of the official goals of the HGP centred on managing scientific work and data, training new scientific specialists, fostering new commercial and profit-oriented relationships between genetics labs and corporate industry, and developing computational and technological tools related to efficiency and cost effectiveness of DNA sequencing, the central goal was to determine the structure, location, and function of the estimated 100,000 human genes (HGP 2000). In essence, the HGP aimed to chart a complete map of the human genome by collecting genetic material, relocating it into one central data bank, reordering the information according to biological classifications, and deciphering the data's relationship to human development. The fundamental basis of the Human Genome Project is the collection of genetic information and the storage of that information in a single and collaborative data bank.

The collection and storage of data that allowed for collaborative and immense work projects were facilitated by the discovery of a technique called sequence tagged sites (STSs) (Olson et al. 1988). Previously, genetic information had been characterized in various organizational formats in a multitude of different laboratories; STS provided a single language for understanding and organizing the data. Moreover, the STS technology allowed for the efficient and cost-effective transfer of genetic information from one laboratory to another and eventually for the centralization of all information: the Human Genome Project (Hilgartner 1995). The HGP reached its goal in June 2000. The completion of the map of the entire human genome was announced as the outcome of a scientific race between corporate scientist and owner of Celera Genomics, J. Craig Venter, and Francis Collins, the director of the public National Institutes of Health's National Human Genome Research Institute. While the media played out the longstanding rivalry between Venter and Collins by focusing on their dynamic and aggressive personalities, which threatened to delay completion of the gene map ("Rivalry" 2000), the larger tensions regarding the social, ethical, and economic implications of the gene map remain. These implications include the dangers associated with genetic determinism and discrimination, genetic engineering and eugenics, as well as corporate and monopolistic control over genetic information and its use.

Genetic determinism is the means by which humanity and human potential are reduced to genetic code. It provides the scientific basis through which people can be identified, rationalized, stigmatized, and controlled through techniques of scientific evaluation. Geneticization is a term applied to the ways in which dominant discourses promote the notion of genetic determinism and reinforce or construct the need to cure genetic traits that are deemed undesirable (Lippman 1991). The concept of eugenics surfaces when techniques of genetic prediction, such as prenatal diagnosis, emerge long before biomedical prevention and cure (if ever they emerge). In the case of prenatal genetic screening of disease for which there is no cure or treatment, fetuses determined to be "at risk" are aborted. Troy Duster (1990) argues that the new eugenic potential is more insidious than old transparent eugenic practices since notions of genetic perfection and imperfection enter ideologically through the back door under the benevolent guise of a curative "genetic fix." Moreover, information acquired from the gene map might not only de-emphasize inquiry into socioeconomic and environmental factors associated with illness and divert funds from preventative health care strategies (McDermott 1998; Paul 1998) but also produce an unemployable and uninsurable biological underclass (Nelkin and Tancredi 1994).

The commercial value of DNA sequences is evident by the extent to which world patent offices are presently congested and staggering under the weight of the many enormous and complex patent applications put forward by researchers who expect to exploit the imminent profitability of human genetic information. However, many argue that DNA, as the building blocks of human life, ought to be used to benefit all humans rather than just those who can afford access to the technology. Dorothy Nelkin and Lawrence Tancredi (1994) contend that the gene map stands to create disparities in ownership, the commercialization of "genetic fixes," and corporate profit, all at the expense of those classified as genetically imperfect or unable to prove their genetic health.

By deconstructing the metaphorical discourse that scientists use to describe their work and to define the Human Genome Project, Mary Rosner and T.R. Johnson (1995) show that scientific claims (to interpret the ultimate book of human DNA, to fix a flawed machine, or to map a mysterious wilderness) are not a reflection of the truth about an objective and natural reality. Rather, scientific claims are interpretive representations of a truth that has been legitimized not only through the already established power of science but also, in this case, through the persuasive metaphors that scientists use to describe them. Using the metaphor that represents nature as a book and the scientist as collector and interpreter, Rosner and Johnson ask what kind of library is being created. More specifically, they wonder what information is unavailable or excluded. What is the nature of the books that make up the library, who are the authors, how are the books printed and bound? What stories do they tell, and who chooses which books will fill the library?

The dubious nature of objectivity in the Human Genome Project is increasingly problematic considering that the meaning of the map has not yet been deciphered. The function of the genes, their relationship to one another, and their relationship to their environment are not clear. As such, the ongoing initiative (as the key to social and scientific progress) is to make sense of the gene map – to make meaning. The gene map is comparable to holding a colossal book with billions of the four letters of genetic code (ATCG) arranged in a certain order but not understanding how the letters interact, how words are formed, and as such what sentences, paragraphs, or chapters mean. Scientists are now attempting to learn the "language of an undiscovered country" ("Learning" 2000), and this process is expected to take several decades (Human Genome Project 2000). Making meaning is never neutral since "meanings are bound up with personal and social ideologies" (Rosner and Johnson 1995, 108). The meaning of the "book of life" will certainly depend, at least in part, on presupposed ideological strategies of classification and interpretation. Even if the scientists agree on the meanings within the DNA text, that does not exclude the possibility that there are not other readings that have not been considered (Rosner and Johnson 1995).

Pharmaceutical and Medical Uses of Biotechnology

Obviously, the Human Genome Project is one of the most significant examples of contemporary biotechnological collection. It gathers, as code, all genetic information in a single human genome for the purpose of using that information for medical, pharmaceutical, and even agricultural applications. As with plant biotechnology, the codification of human life is highly involved with those interested in commercial exploitation. Both plant and animal genetic information is a highly prized resource in the burgeoning pharmaceutical and medical industries. Also, as with plant biotechnology, this type of scientific and medical development and the associated applications raise considerable public debate and resistance in the face of great promises of disease eradication and social progress.

Much of medical research is now qualified by genetics in terms of both diagnosis and treatment. This radically new approach in medicine has meant collection of information about the genetic makeup and function of the human body. It also entails genetic screening as the new determination of difference as those with and those without disease or the risk of illness. In the early modern collections of natural science, difference was determined by relatively gross features: colour, size, obvious physical attributes.

Behaviour, another early delineator of difference and signifier of valuable addition to a collection, was also considered as that of the entire being. Biotechnology in the medical sciences and the pharmaceutical industry reduces behaviour to genetic expression and difference to microscopic features.

Reports of medical discoveries regarding some of our most threatening illnesses fill newspapers almost daily. Following the successful cloning of Dolly the sheep came Polly the sheep, genetically altered to carry the human gene responsible for producing the blood-clotting material Factor IX ("Polly" 1997). This blood product is what hemophiliacs lack and try to replace with human-derived serum. This traditional treatment of hemophiliacs has also led to grave contamination situations when blood and its products in Canada were tainted with the HIV virus. So genetic engineering of animals to provide the serum solves a serious social risk. Medical gene hunters have also "tracked down" the genetic link to so-called good cholesterol ("Genetic Link" 1999). This link paves the way to treating heart disease, one of the most common causes of death in developed countries. Also in regard to the treatment of heart disease, French medical scientists have spliced the human gene responsible for the production of apolipoproteinE (apoE) into mice to effectively reduce artery-clogging low-density lipoprotein ("Science Spectrum" 2000). Other medical researchers are "exploring agents" and the gene that preclude smokers, especially women, to lung cancer ("Gene" 2000). They found that nicotine stimulates the GRPR gene in lung cells, where it is usually not active. This gene is seen as being responsible for supporting a protein growth factor that leads to cancerous cell proliferation. In Canada, a small genetic flaw that is responsible for a debilitating neurodegenerative disease named for and affecting primarily French Canadians from northeastern Quebec, spastic ataxia of Charlevoix-Saguenay (SACS), has been discovered ("One Lousy Mistake" 2000). Finally, the director general of Britain's Cancer Research Campaign, Gordon McVie, promises that "we are on the eve of a genetic revolution which shall see a leap in the cure rates for cancer and a whole range of new, effective treatments" ("Cancer" 2000).

With the post-Linnean genetic codification of life comes a level of financial exploitation of natural discoveries and the reorientation of nature never seen before. Patenting and property rights are key to the burgeoning business of biotechnical applications in medicine. Pharmaceuticals and biotechnology firms are the main vehicles for the transfer of nature as information into profit. Patenting of plant life has been extended to the animal realm and includes human genes in many developed nations, including the United States and Canada. The commercial expansion in medical biotechnology and pharmaceuticals has been so remarkable that it has spawned new models of business. FIPCOs, or fully integrated pharmaceutical companies, refer to a vertical company structure that carries on activities ranging from research to the market. This model has been replaced by the virtually integrated pharmaceutical company (VIPCO) that specializes in one area of business, such as "drug discovery," and farms out all other aspects of the commercialization process. There are also "platform" and "integrated platform" models that focus on a single set or related sets of technologies respectively. CBC Radio reported that Stockwell Day made Preston Manning the science and technology critic in Day's shadow cabinet in order to address Canada's financial competitiveness in a globalized economy (CBC Radio 2001). This link between science, technology, and business is also reflected in Queen's University's private MBA program, which only accepts applicants with a background in science and engineering.

The future holds "model-free" company organization for biotechnology where R&D is the only common feature (Hodgson 2000, 31). Or, in other words, as long as the bottom line is safeguarded, anything goes. And it can go a long way. The Australian biotechnology company Autogen has secured the exclusive rights to the gene pool of the entire population of Tonga to develop treatments for cancer and heart disease. DeCode of Iceland collated the available genealogical, genetic, and medical data on the island's 290,000 people and then sold exclusive rights to the population's genetic information for twelve years to the Swiss pharmaceutical Hoffmann-LaRoche in 1998 for $200 million (see Eischen, this volume, Chapter 6, for a detailed analysis of deCode). This led to the discovery of a genetic link to osteoporosis and enormous potential revenues for LaRoche ("Biotech Company" 2000). Relatively closed and small communities such as Tonga and Iceland are a valuable resource in genetic trawling as genetic-linked illnesses can be more easily traced through reproductive lines, which can lead to a much quicker determination of the gene linked to the particular ailment. Quicker discoveries spell less investment in R&D and higher profits. There is also the dark and hardly publicized side of collecting genetic communities – genetic science can also be used to develop racially specific "biochemical weapons" ("Crop Circles" 1998).

This level of commercial exploitation in biotechnology has not escaped public notice. As a reaction to public outcries against the buying and selling of animal, especially human, life, many large biotechnology and pharmaceutical companies now mount extensive public relations campaigns. Some have even been forced to make their genetic collections public, as with the Human Genome Project. One of the world's largest pharmaceuticals, Glaxo-Wellcome, has established the Wellcome Trust to fund a substantial public relations enterprise that includes an information service and an art and science gallery in central London, England. The information made available to the public is not raw genetic data (as with the Human Genome Project) but information about medical science and its community. The service provides a database of science policies, ethical issues, and employment and funding information. The gallery, Two10, includes traditional representations of medical life, such as the painting *Concentration – Operating Theatre* (Powell 1995). It also has remarkably enlarged medical images such as the colour-enhanced photo of a single hair shaft and electron microscopic representations of cells and viruses. The pharmaceutical's glossy newsletter highlights the latest in biotechnological research projects, especially those funded by the company. All these reports are presented in a highly accessible and pleasing photo format with little contention or debate of biotechnology issues (*Wellcome News* 1998).

The need for such public relations measures can be explained by recent moves by pharmaceuticals to protect their investments and profits in the face of the pandemic AIDS. More than fourteen million Africans have died of AIDS since the 1980s, and researchers say that this is only the beginning. The effects of the disease are devastating social and economic systems as death rates leave millions of children infected orphans and as unemployment soars ("Dire Time" 2000). Glaxo-Wellcome moved to block the importation of cheap versions of one of its AIDS therapy drugs, Combivir made in India by Cipla, from going to Ghana. Another pharmaceutical, Pfizer, was forced to provide its drug Fluconazole (used to treat AIDS-related infections) to South Africa for free when campaigners started bringing in another, much cheaper version ("Glaxo" 2000). The activist medical charity Médecins sans frontières has been one of the key players in bringing affordable AIDS drugs to Africa. It points out how even arguments that pharmaceuticals require patents to protect their investments wear thin when one considers that the Combivir drugs are old and the company has already more than received its return on them ("Glaxo" 2000).

Conclusion

Many believe that biotechnology represents a brave new world of scientific innovation, but biotechnology as a complex social initiative follows the problematic principles of collection established over 300 years ago. Predominantly, biotechnology represents the appropriation of foreign and exotic entities as genetic information into a dominant culture's collection. There that information is encoded with the prime intent to protect dominant interests in what is collected. This interest is clearly commercial. Linnean classification remains the basic term of reference and a way of legitimizing knowledge of the biotech world. However, unlike that of the past, this knowledge is clearly marked for direct commercial exploitation by those who appropriate the information entities from their natural and cultural homes. Early naturalists collected samples (cultural as well as natural) from foreign shores to be categorized and displayed in early museums. The collected entities were made "exotic" in this process and valuable in the sense that they contributed to "libraries" and classification schemes of the dominant culture. Exactly the same occurs within contemporary biotechnology as the look of life becomes the book of life: what was once determined as both different and collectable by gross appearances is now considered in the same way but according to microscopic genetic code. Display continues, but beyond the societies of European gentlemen scientists and art collectors and within global markets where steep admission charges apply.

In the context of the Human Genome Project, the genetic display is not yet fully realized, and the mounting of the exhibit does not go without serious protest. Many do not seek admission to the new museum of humanity since they resist the colonizing effects that this type of exhibit represents. Increasingly, however, with globalization and monopolization of information of all kinds, peoples are forced to negotiate according to dominant interests. There is an expectation that all peoples accept the language of life described by these collections, but this is not always the case. There are extensive and well-organized pockets of resistance to what is determined as significant and inevitable by global corporate curators.

References

Biography. 2000. Discovery Channel, 29 December.

Biopat. 2001. "Kerfersteinia, Your Choice: Name a Frog or an Orchid!" On-line at <http://www.gtz.de/biopat/english/eng-biopat-tour1.htm>, retrieved 26 November 2001.

"Biotech Company Finds Rich Pickings in South Pacific Gene Pool." 2000. *Guardian Weekly,* 7-13 December, 34.

"Cancer Will Be Beaten within 50 Years, Expert Says." 2000. *National Post,* 5 January, A14.

CBC Radio. 2001. *CBC Radio News,* 5 January.

Clifford, J. 1998. "On Collecting Art and Culture." In Nicholas Mirzoeff, ed., *Visual Culture Reader,* 94-107. New York: Routledge.

"Clone Hope to Save Panda." 1999. BBC News, 24 December. On-line at <http://news.bbc.co.uk/>, retrieved 11 January 2001.

"Clone Plan for Extinct Goat." 2000. BBC News, 11 January. On-line at <http://news.bbc.co.uk/>, retrieved 11 January 2001.

"Cloned Ox Dies of Natural Causes." 2001. CBS News, 12 January. On-line at <http://www.cbsnews.com/stories/2001/01/12/tech/main263745.shtml>.

Coen, E. 1999. *The Art of Genes.* New York: Oxford University Press. On-line at <http://www.educationplanet.com>.

"Crop Circles." 1998. *Guardian Weekly,* 15 July, 4.

de la Perrière, Brac, Robert Ali, and Frank Seuret. 2000. *Brave New Seeds: The Threat of Transgenic Crops to Farmers in the South.* Toronto: Fernwood Books.

Denton, M. 1996. *Evolution: A Theory in Crisis.* 2nd ed. New York: Adler and Adler.

Department of Energy and the National Institutes of Health. 1991. *Understanding Our Genetic Inheritance: The US Human Genome Project: The First Five Years, 1991-1995.* Washington, DC: NIH Publication.

"A Dire Time." 2000. *Toronto Star,* 16 July, B1.

Duncan, C. 1998. "The Modern Art Museum." In Nicholas Mirzoeff, ed., *Visual Culture Reader,* 85-93. New York: Routledge.

Duster, T. 1990. *Backdoor to Eugenics.* London: Routledge.

"Endangered Species in Your State." 2000. On-line at <http://www.endangeredspecie.com>, retrieved 9 January 2001.

"Facts about Endangered Species." 2000. On-line at <http://www.endangeredspecie.com>, retrieved 9 January 2001.

Findlen, P. 1996. *Possessing Nature: Museums, Collecting, and Scientific Culture in Early Modern Italy.* Berkeley: University of California Press.
Fiore, K.H., ed. 1997. *A Guide to the Borghese Gallery.* Rome: Edizione de Luca.
Fusco, C. 1998. "The Other History of Intercultural Performance." In Nicholas Mirzoeff, ed., *Visual Culture Reader,* 363-71. New York: Routledge.
Galton, F. 1909. *Essays in Eugenics.* London: Eugenics Education Society.
Gardner, H. *Art through the Ages.* 1975. 6th ed. Revised by Horst de la Croix and Richard G. Tansey. New York: Harcourt, Brace, Jovanovich.
Gene Letter. 2000. On-line at <http://www.geneletter.com>.
"Gene Raises Female Smokers' Risk." 2000. *National Post,* 5 January, A15.
"Genetic Link to Good Cholesterol Found." 1999. *Globe and Mail,* 3 August, A2.
Geno-Types List. 1998. "US Patents on New Genetic Technology Will Prevent Farmers from Saving Seed." On-line at <http://www.rafi.ca>.
"Glaxo Stops Africans Buying Cheap AIDS Drugs." 2000. *Guardian Weekly,* 7-13 December, 7.
Gostin, L. 1994. "Genetic Discrimination." In Robert Weir, Susan Lawrence, and Evan Fales, eds., *Genes and Human Self-Knowledge,* 122-63. Iowa City: University of Iowa Press.
Hilgartner, S. 1995. "The Human Genome Project." In Sheila Janasoff et al., eds., *Handbook of Science and Technology Studies,* 302-16. Albany: Sage Publications.
Hodgson, J. 2000. "Crystal Gazing the New Biotechnologies." *Nature Biotechnology* 18 (1): 29-31.
"How Noah Could Clone a New Ark." 2001. *Guardian Unlimited,* 7 January. On-line at <http://www.observer.guardian.co.uk>, retrieved 11 January 2001.
Human Genome Project. 2000. HGP website. On-line at <http://www.ornl.gov/TechResources/Human_Genome/home.html>, retrieved 20 December 2000.
"India to Clone Cheetah." 2000. BBC News, 16 October. On-line at <http://news.bbc.co.uk/>, retrieved 11 January 2001.
International Center for Agricultural Research in the Dry Areas. 2001. On-line at <http://www.icarda.cgiar.org/>, retrieved 11 June 2001.
International Plant Genetic Resources Institute. 2000. On-line at <http://www.ipgri.cgiar.org/>, retrieved 11 June 2001.
Kevles, D. 1985. *In the Name of Eugenics: Genetics and the Uses of Human Heredity.* New York: Alfred A. Knopf; Cambridge, MA: Harvard University Press.
"Learning the Language of Undiscovered Country." 2000. *National Post,* 13 March, A11.
Lee, T. 1991. *The Human Genome Project: Cracking the Genetic Code of Life.* New York: Plenum Press.
Lippman, A. 1991. "Prenatal Genetic Testing and Screening: Constructing Needs and Reinforcing Inequalities." *American Journal of Law and Medicine* 17: 15-50.
McDermott, R. 1998. "Ethics, Epidemiology, and the Thrifty Gene: Biological Determinism as a Health Hazard." *Social Science and Medicine* 47: 1189-95.
McLaren, A. 1990. *Our Own Master Race: Eugenics in Canada, 1885-1945.* Toronto: McClelland and Stewart.
Madeley, J. 2000. *Hungry for Trade: Does Trade Help or Hinder Food Security?* Toronto: Fernwood.
Manicad, G. 1996. "Biodiversity Conservation and Development: The Collaboration of Formal and Non-Formal Institutions." *Biotechnology and Development Monitor* 26: 15-17.
"Map of Rice Genome Sows Seeds of Change." 2000. *Toronto Star,* 23 April, B6.
"Monkey Born with Genetically Engineered Cells." 2001. *New York Times,* 12 January, National Desk Section. On-line at <http://college1.nytimes.com/guests/articles/2001/01/12/640698.xml>, retrieved 16 January 2001.
National Council of Welfare. 2000. *Justice and the Poor.* Ottawa: National Council of Welfare.
National Wildlife Federation. 2000. On-line at <http://www.nwf.org>, retrieved 11 January 2001.
Nelkin, D. 1993. "The Social Power of Genetic Information." In D. Kevles and L. Hood, eds., *The Code of Codes: Scientific and Social Issues in the Human Genome Project,* 177-90. Cambridge, MA: Harvard University Press.
Olson, M., L. Hood, C. Cantor, and D. Botstein. 1988. "A Common Language for the Physical Mapping of the Human Genome." *Science* 245: 1434-35.

"One Lousy Mistake." 2000. *National Post,* 1 February, A17.

Oxfam Canada, Greenpeace, and CUSO. 2000. Press release, Ottawa, 11 October.

Paskey, J. 2000. "Research Deal Drives a Wedge at University in Canada's Rural Heartland." *Chronicle of Higher Education*, on-line at <http:chronicle.com/weekly/v46/i32a07501.htm>.

Paul, D. 1998. *The Politics of Heredity: Essays on Eugenics, Biomedicine, and the Nature-Nurture Debate.* New York: SUNY Press.

Pollan, Michael. 1994. "The Seed Conspiracy." *New York Times Magazine,* 20 March, 49-50.

"Polly, the First Designer Clone, Engineered to Help Hemophiliacs." 1997. *Toronto Star,* 19 December, A3.

Powell, V. 1995. *Concentration – Operating Theatre, Painting.* London: Two10 Gallery.

"The Race to Clone the First Human." 1998. BBC News, 26 August. On-line at <http://news.bbc.co.uk>, retrieved 15 January 2001.

"Rivalry Could Stall Genetic Breakthrough." 2000. *National Post,* 13 March, A1.

Rosner, M., and T.R. Johnson. 1995. "Telling Stories: Metaphors of the Human Genome Project." *Hypatia* 10 (4): 104-29.

Schmidt, P. 2000. "Public Universities Get Money to Attract High-Tech Industry." *Chronicle of Higher Education,* 15 February. On-line at <http://chronicle.com/weekly/v46/i25/25a04201.htm>.

"Science Spectrum: From Gene to Heart." 2000. *Queen's Journal,* 15 February, 13.

"Scientists Make Human Clone Claim." 1998. BBC News, 16 December. On-line at <http://news.bbc.co.uk/>, retrieved 15 January 2001.

Shiva, V. 1997. *Monocultures of the Mind: Perspectives on Biodiversity and Biotechnology.* 3rd ed. London: Zed Books.

Stewart, S. 1985. *On Longing: Narratives of the Miniature, the Gigantic, the Souvenir, the Collection.* Baltimore: Johns Hopkins University Press.

Suzuki, D. 1999-2000. "Pic Press Presents a Three-Part Series on Genetically Modifying Our Food." *Progressive Independent Community Press,* December-January, 11.

"Tasmanian Tiger May Growl Again." 1999. BBC News, 14 May. On-line at <http://news.bbc.co.uk/>, retrieved 11 January 2001.

"Taxonomy." 1985. *Encyclopedia Britannica* 14: 939-45.

UNEP/GEF Biodiversity Planning Support Program. 2000. Listserv notice, Nairobi, 27 November.

Wellcome News. 1998. 14, Q1.

West, D. 1994. "Why I Don't Like Museums: A Reply to the Commentary 'Personal, Academic, and Institutional Perspectives on Museums and First Nations' by Robert R. James." *Canadian Journal of Native Studies* 14 (1): 363-68.

Wildlife Conservation Society. 2000a. On-line at <http://wcs.org>, retrieved 11 January 2001.

–. 2000b. "Campaign to Protect Critically Endangered Beluga Sturgeon and Other Threatened Sturgeon Species." On-line at <http://wcs.org/3422?newsarticle=7822g>, retrieved 8 December 2001.

World Resource Institute. 2000. On-line at <http://www.wri.org/wri/wri.html>, retrieved 11 June 2001.

World Wildlife Fund. 2000. On-line at <http://www.worldwildlife.org>, retrieved 15 January 2001.

"You, Too, Can Buy a Species." 2000. *Toronto Star,* 5 March, F8.

9
The Production, Diffusion, and Use of Knowledge in Biotechnology: The Discovery of BRCA1 and BRCA2 Genes

Robert Dalpé, Louise Bouchard, and Daniel Ducharme

Watson and Crick's 1953 revelation on the structure of DNA represents one of the past century's most important scientific breakthroughs (Heilbron and Bynum 2000). Placing emphasis on the potential impacts of genetic research on disease detection and medical treatment, molecular biology, as an academic discipline, developed rapidly, and its scientific community benefited significantly (Kenney 1986). In 1973, Boyer and Cohen's discovery of the recombinant DNA technique led to the isolation of genes and their transfer from one organism to another. In 1975, Milstein and Kohler introduced the technique of monoclonal antibodies.[1] Both had direct industrial applications, thus opening the new field of genetic engineering and biotechnology.[2] This in itself constitutes a turning point that ushered in the commercial exploitation of pure research. Stanford University and the University of California obtained three patents protecting Cohen and Boyer's recombinant DNA technique, while a new firm, Genentech (cofounded by Boyer) was created to commercialize it. Both universities would make over $200 million US in royalties during the lifetime of these patents, and Genentech would emerge as a leader in biotechnology.[3] New firms, many of them founded by university researchers, would subsequently be created to develop commercial applications derived from this kind of research.

Due to potential commercial applications of a large segment of genetics and biotechnology research in public organizations, researchers established close interactions with industry, and new practices were set. For instance, following Boyer's lead, several university researchers created firms and managed them while retaining their academic positions. University-based biotechnology research is therefore more largely financed by private contracts, and patenting is more frequent than in other disciplines. The involvement of industry, however, imposes new constraints on scientific research, for instance in the area of information diffusion since industrial commercialization requires that firms appropriate research results and secure property rights prior to public disclosure (Blumenthal et al. 1997; Powell and Owen-Smith 1998). The current public debate over intellectual property derived from the Human Genome Project is perhaps the best illustration of this happening (Buttler 2000; Lewis 2000).

This chapter deals with the new research dynamics in biotechnology generated by industry's direct relations with university and public laboratory researchers. Our objective is to determine how researchers act in and respond to their new environments and assess constraints imposed. We will present results from a case study dealing with the discovery of two genes associated with breast and ovarian cancer, BRCA1 and BRCA2. Localization of the first gene, BRCA1, in 1990, and its identification in 1994, represented major scientific events because it was the first gene discovered for a widespread and serious disease. Starting in the mid-1990s, industrial applications dealt with the development of genetic testing for the detection of gene mutations. Industry continues to be a major actor in gene discovery through in-house research, financing of university researchers, and as patent assignee. The industrial leader, Myriad Genetics of Salt Lake City, was in the race at each stage and gained priority for BRCA1 identification and BRCA2 localization. As we will see later, firms and public organizations engaged in frequent conflicts concerning patenting of both genes and their subsequent industrial applications (Kotulak 1999; Meek 2000; UK Clinical Molecular Genetics Society 1999).

The next section presents our main conclusions from studies examining the new research dynamic in public research organizations, followed by a description of breast and ovarian cancer genes research since the early 1990s aimed at a better understanding of researchers' environments. This section is based on a bibliometric study of the evolution of scientific publications dealing with BRCA1 and BRCA2 genes. Finally, the strategies of researchers are analyzed based on in-depth interviews of seven researchers who represent different profiles in research dynamics.

The Dynamic of Biotechnology Research

At the end of the Second World War in the United States, scientific researchers helped to convince the federal government of the political, social, and economic advantages of large research investments (Bush 1980). The 1950s and 1960s were characterized by rapid increases in public funding, while private financing, although sizable during the first half of the century, grew only slightly. In the implementation of this research dynamic known as the "Republic of Science,"[4] the research lobby emphasized the impact of new scientific knowledge on the recent Allied victory and on the welfare of populations (e.g., the effect of health research on life expectancy) (Ruivo 1994). According to the "linear model" underlying this research dynamic and its interaction with society, university research increases the knowledge stock on which industry will draw to develop new technologies. Through research councils, researchers themselves exercise control, and funds are allocated on the basis of peer review, according to the scientific merits of projects.

Several studies in the sociology of science maintain that a new research dynamic is progressively replacing the Republic of Science. According to Gibbons et al. (1994), the Republic of Science, or Mode 1, characterizing curiosity-driven and disciplinary research, is surpassed by Mode 2, dealing with transdisciplinary and problem-driven research structured around knowledge application. Furthermore, in Mode 2, research is carried out simultaneously in a variety of public and private organizations. The "Triple-Helix" model places emphasis on institutional arrangements and communication networks linking university, government, and industry researchers (Etzkowitz and Leydesdorff 1997).[5] In this model as well, new actors are involved, and organizational frontiers are blurred. Both models are based on two assumptions. Scientific research becomes more beneficial to industrial innovation if knowledge is properly diffused to industry, and the rate of success depends on frequent interaction between university and other research organizations (Edquist 1997). The assumptions behind the classic linear model are questioned. By the same token, Slaughter and Leslie (1997) maintain that such direct interaction implies a change in university culture and the implementation of new management practices modelled on rules and norms internal to industry. This more direct interaction is characterized by new rules governing the allocation of funds, making them more attuned to market transactions, and by a greater role of industry in research.

Such studies do not sufficiently take into account the diversity of academic research practices (Dalpé and Ippersiel 2000). For example, more applied disciplines, such as metallurgy and chemical engineering, emerged in the nineteenth century in close collaboration with industry and maintained such interaction even during the golden age of the Republic of Science (Rosenberg and Nelson 1994). In this respect, the Republic of Science characterizes well the discourse of researchers and research council policies, but it fails to properly reflect the heterogeneity of research practices in universities (Godin 1998).

What characterizes these new research practices involving a more direct interaction between university researchers and industry is that now it affects specific academic disciplines. The share of private funding of university R&D increased from 4 percent in 1980 to 7 percent for the United States and to 11 percent for Canada in 1996 (Dalpé and Ippersiel 2000). Information technology and biotechnology, however, constitute the largest share of private funding. Statistics on university-industry interactions, such as contracts, patents, and licences, show that in large American universities this increase is mainly attributable to the commercialization of biotechnology (Mowery et al. 2001). Studies in sociology and economics dealing with biotechnology maintain that the interaction between university and industry draws its specificity from three major factors.

First, the creation of new firms commercializing biotechnology stems directly from new knowledge developed in universities (Kenney 1986). The diffusion of knowledge generally takes place primarily through indirect mechanisms, such as human resources training or the increase of the knowledge stock available to firms. For biotechnology, diffusion is done by researchers themselves through the creation of new firms to commercialize this knowledge. Existing companies, such as large pharmaceutical firms traditionally staffed with biochemists, were unable to exploit new biotechnology knowledge without the direct interaction with biotechnology researchers. In that respect, diffusion of knowledge in biotechnology, because of its novelty and complexity, cannot take place rapidly and fully without contact between researchers and industry representatives (Argyres and Liebeskind 1998; Zucker, Darby, and Brewer 1998). By the same token, this explains why biotechnology firms in North America tend to form clusters around leading public research centres: diffusion of knowledge characterized by tacit knowledge demands proximity (Audretsch and Stephan 1996; Porter 1998).

Second, institutional arrangements largely specific to biotechnology were implemented to allow closer relations between university researchers and industry. The most interesting arrangement is that university researchers were allowed to keep their academic positions while holding office in their firms, as scientific advisors, for instance. This institutional arrangement distinguishes biotechnology firms from information technology firms, where researchers rapidly left universities upon founding their firms (Kenney 1986; Prevezer 1997). In most European countries, researchers are civil servants, and direct involvement in a firm is illegal (Giesecke 2000; Senker and Van Vliet 1998). Various other institutional arrangements also apply in other research areas but with a greater importance in the case of biotechnology. In the 1980s, university-industry liaison offices were established in most universities. The implementation of the Bayh-Dole Act in 1980 to award intellectual property to universities in projects funded by research councils helped to foster commercialization (Shohet and Prevezer 1996). Finally, new rules in the United States on patent policy were crucial in securing property rights in biotechnology, which is a prerequisite to obtaining venture capital.

Third, the performance of biotechnology firms is explained by their integration with large research and industrial networks (Powell et al. 1999). Concerning the former, a firm's success in the stock market depends on its links with leading researchers (Deeds, Decarolis, and Coombs 1997). For industrial networks, the support of other actors, such as large pharmaceutical firms and venture capital, also depends on the quality of the scientific research. In these large networks, intellectual property is a crucial issue. According to transaction cost economics, these networks are efficient only if formal rules are enforced to control the diffusion of information (Kogut 1989). Patenting new knowledge is required but does not completely deter imitation by competitors. For its part, the sociology of networks places emphasis on the social dimensions of these interactions and more specifically on trust between actors (Granovetter 1985). Face-to-face contacts and long-term relationships solidify networks and deter opportunistic behaviour. Financial incentives, such as a share of capital in the firm, strengthen the loyalty of researchers.

These various approaches do not dwell sufficiently on the impact of such direct interactions with academic research practices. Powell and Owen-Smith argue that the most severe side effects are "increased politicization of government research funding, a growing winner-take-all contest between the have and have-not universities, and subtle but potentially profound changes in the culture of academic research" (1998, 253). The importance of private funding and the benefits that researchers and their university derive from commercialization might affect the dynamics of research practices themselves in at least three ways. First, university researchers with private-sector support and better resources may have an advantage over competitors for granting council funds. Second, a large number of academic biotechnology researchers have established relationships with firms (Zucker, Darby, and Brewer 1998). Third, the net result of these developments is that frontiers between public and private organizations are increasingly blurred.

The impact most frequently studied is that of the new rules concerning intellectual property. Economists believe that university knowledge can be commercialized only if firms can appropriate it (Pisano, Shan, and Teece 1988). New rules governing the publication of research results and diffusion of information in scientific journals therefore tend to be more frequently delayed when university researchers are in interaction with firms and benefit from private funding (Blumenthal et al. 1997). Researchers' autonomy and probity can be questioned when they are largely funded by a given firm, with which they can be in conflict of interest (Cho et al. 2000; Powell and Owen-Smith 1998).

Research on BRCA1 and BRCA2 Genes[6]

Breast and ovarian cancer gene research is frequently used as an illustration of this new dynamic of research and its side effects (Cassier and Foray 1999). Most performing teams would be those obtaining industrial support and those building large transdisciplinary and interorganizational networks (Powell and Owen-Smith 1998). As for intellectual property, researchers discuss the social benefits of patenting and exclusive licensing on research per se as well as on health costs.

Two genes associated with breast and ovarian cancer have been identified. Mutations of these genes are linked with about 5 percent of breast cancer cases (Arver et al. 2000). Two data sources are exploited to describe the evolution of research on these genes during the past ten years. The first one is a search of the medical press that followed these discoveries step by step. The second source is a bibliometric study of scientific publications on both genes. Bibliometrics is devoted to the quantitative analysis of documents, such as papers or patents, in order to describe the dynamics and evolution of a research field (Gauthier 1998; Van Raan 1988). The database includes 803 scientific documents published between 1993 and 1998.[7] In order to obtain a profile of research on these genes, information gathered concerns authorship, institutional affiliations, and citations appearing in these publications. A special emphasis was put on collaboration, as measured by coauthorship for publications having authors with different institutional affiliations.

The BRCA1 gene, localized on chromosome 17q, is believed to function as a tumour suppressor gene, and mutations increase the risk of breast and ovarian cancer (Elwood 1999). More than 350 different mutations have already been identified, and only a few have been found in multiple families. Furthermore, this gene is much larger than previously isolated genes. In this context, the identification of germline mutations often requires the complex task of sequencing the DNA throughout the gene. Estimates claim that women carrying a BRCA1 mutation have a 56-87 percent risk of developing breast cancer and a 20-60 percent risk of developing ovarian cancer during their lifetimes (Armstrong, Eisen, and Weber 2000; Struewing et al. 1997). These differences in percentage show that the various teams still have divergent points of view concerning the impact of these mutations.

In the late 1980s, several North American and European teams, while trying to identify different genes associated with various diseases such as cystic fibrosis or sickle-cell anemia, were looking for genes related to breast cancer. The identification of BRCA1 was achieved by a team led by Mary-Claire King in December 1990 (Hall et al. 1990). This important scientific breakthrough launched an international race for sequencing the gene (Davies and White 1996). The competing teams met regularly under the International Breast Cancer Linkage Consortium, where members exchanged data and shared genetic databases of families suffering from hereditary breast and ovarian cancer. The 1993 meeting offered the first estimate of the impact of gene mutations on cancer (Easton et al. 1993).

The gene BRCA1 was finally sequenced by a team led by researchers from the University of Utah and the Salt Lake City–based firm Myriad Genetics, cofounded by Mark Skolnick, from and still affiliated with the University of Utah. The paper published in *Science* in October 1994 was coauthored by researchers from three other organizations, the NIH Institute of Environmental Health Sciences, McGill University, and the pharmaceutical firm Eli Lilly (Miki et al. 1994). Competition was so strong that the paper was submitted to *Science* without the sequence.[8] The release was a major media event. The initial announcement was made on NBC News, and *Science* indicated that the paper was accepted for publication and atypically sent it directly to the press the next day, even prior to its publication in the journal.

Conflicts about intellectual property started almost immediately. The first was between the University of Utah and Myriad Genetics and the NIH. Even if two researchers from NIH were coauthors of the paper, their names were excluded as inventors when Myriad Genetics filed for a patent. According to Myriad managers, the scientific and financial contribution of NIH was rather small, the project being financed by the company's own funds and by the large pharmaceutical firm Eli Lilly. In February 1995, an agreement was reached to add NIH researchers and to share royalties. The second conflict emerged in 1996 between Myriad Genetics and its competitor Oncormed. The previous conflict delayed patenting for Myriad Genetics, so that Oncormed won the first patent delivered in the United States. This second conflict was resolved in May 1998 when Oncormed, suffering from severe financial problems, sold its rights to breast and ovarian cancer gene patents to Myriad. A third, broader debate developed over the very possibility of patenting human genes. Among others, representatives of various women's health groups and coalitions opposed to gene research were also active in questioning the social benefits of patenting genes (Oberbeck 1996). Their arguments addressed the very idea of patenting human characteristics and the possibility of discrimination by employers and insurers should the results of tests be revealed.

The second gene, BRCA2, localized on chromosome 13q, is twice the size of BRCA1. More than 100 mutations have been identified, and only a few have been observed in multiple families. Women carrying a BRCA2 mutation have a 56-87 percent risk of developing breast cancer by age eighty and a 10-20 percent risk of developing ovarian cancer during their lifetimes.

The announcement of BRCA2 localization coincided with the sequencing of BRCA1 in September 1994. Furthermore, there was overlap between both research teams since researchers from Myriad Genetics and the University of Utah were involved in both projects. The paper was led by David Goldgar from the University of Utah and included thirty-one coauthors from six countries, some of whom were from the British Institute of Cancer Research (Wooster et al. 1994). Priority for BRCA2 sequencing was much more contested. Due to Myriad Genetics' strategy concerning intellectual property on BRCA1, relations between various research groups were difficult, and the team finally collapsed (Butler and Gershon 1994). A consortium led by Mark Stratton of the British Institute of Cancer Research, including thirty-eight coauthors from six countries, published its results in *Nature* in December 1995, without the participation of Myriad Genetics or the University of Utah (Wooster et al. 1995). The day before the publication, Myriad Genetics disseminated a press release indicating that the company had submitted the complete sequence of BRCA2 to GenBank and applied for a patent. Its own paper was published in March 1996, mentioning that Stratton's paper "reported a partial sequence and six mutations" (Tavtigian et al. 1996, 333). In order to avoid conflicts with organizations such as NIH, Myriad's strategy was to finance sequencing only through private funds. The British group also applied for a patent in the United Kingdom, and a licence was awarded to Oncormed. Myriad Genetics also recovered the rights when it acquired some of Oncormed's assets.

Up to 1996, the main research subject was the identification and localization of genes and mutations. Interactions between researchers having the largest number of publications in our database are described in Figure 9.1, generated by factor analysis of the matrix of coauthorship.[9] Six networks are clearly identified.

The first network includes UK researchers, mainly from the British Institute of Cancer Research and from Cambridge University, as well as groups

Figure 9.1

Research teams

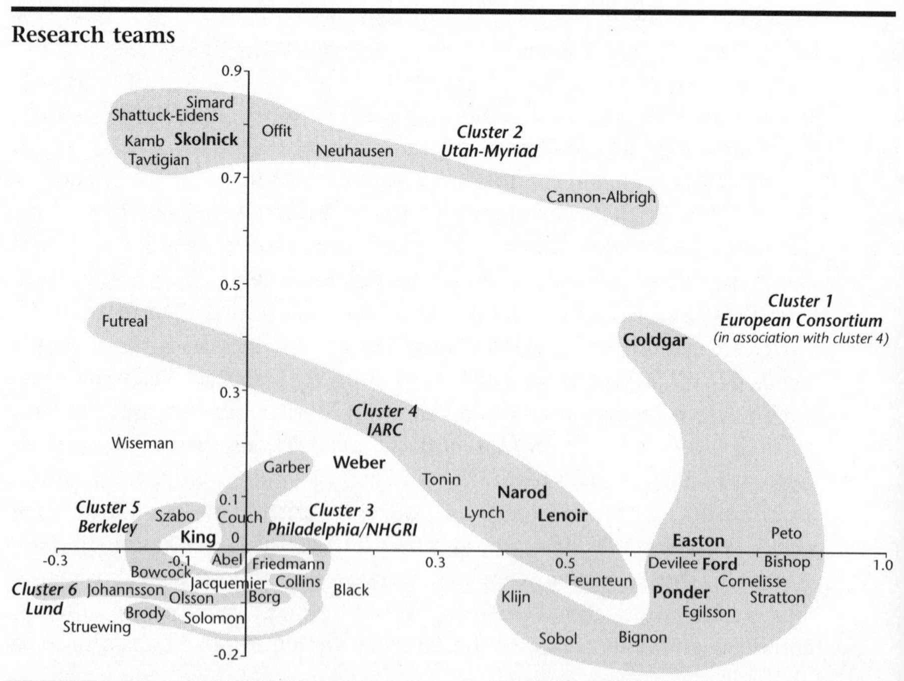

from France and the Netherlands. It represents a rather diversified group, which together with the fourth network, around the French International Agency for Research on Cancer (IARC), were the leaders in the International Breast Cancer Linkage Consortium. The second network is much more homogeneous; with only two exceptions, it consists of researchers from Myriad Genetics and the University of Utah. This second network is the only one to integrate researchers from industry and, when compared with the first, is much tighter. It does, however, have some interactions with leading researchers from other networks. The third network includes researchers from the University of Pennsylvania and from the National Human Genome Research Institute (NHGRI). The fifth network includes Mary-Claire King's former collaborators. Having interactions with other teams, she does not fully participate in the fifth network. The sixth network brings together three researchers from the University of Lund.

This network analysis shows greater interactions between researchers and the role of American, British, and French organizations in structuring these interactions. Obviously, fewer interactions took place among the six networks. A small group of researchers, such as David Goldgar, Barbara Weber, and Steven Narod, do have some links with almost all networks.[10] From the late 1980s up to 1996, two objectives strengthened collaboration between research groups. The first one was to share databases, each team having only a small number of families and often being from the same geographical area. The International Breast Cancer Linkage Consortium played an important role here. The second objective, which will be discussed more fully in the section based on interviews, was to convince the broader medical community of the value of such research results and of their potential impacts on cancer treatment and broader public health issues, such as the possibility of implementing national programs of genetic screening.

If researchers collaborate for the advancement of science, activities related to commercialization are more conflictual. Tensions between Myriad Genetics, NIH, Oncormed, and the British Institute of Cancer Research have already been mentioned. More recently, Myriad Genetics tried to enforce its patents by requesting that other research groups pay royalties whenever testing was done on a commercial basis (Hurst 2001; Kotulak 1999). Researchers from the University of Pennsylvania and the British Institute of Cancer Research have publicly criticized Myriad Genetics for this action, stating that it will increase research and genetic screening costs (Kotulak 1999; Meek 2000). In September 2001, the French Institut Curie, supported by its minister of research, announced its decision to challenge a recent Myriad European patent (Butler and Goodman 2001). A few weeks later, the Ontario minister of health allowed his hospitals to develop their own test and to perform it without paying royalties to Myriad, in spite of potential legal retaliations (Abraham 2003; Benzie 2001).

Although Myriad Genetics established its property rights on both genes, sales of its genetic tests are still very weak. Several factors explain the slow diffusion of this new technology. First, scientific uncertainty remains concerning the identification of mutations and their frequency, distribution, and predisposition to disease. A French group recently argued that Myriad's tests fail to detect a large number of mutations (Gad et al. 2001). Since 1996, research on both genes has been devoted to their function and to the understanding of their impacts on cancer (Scully and Livingston 2000). The second factor deals with clinical applications. Flaws in the performance of these tests, their capacity to predict future disease, and their interpretation, when added to genetic screening costs, represent an additional obstacle.[11] In the broader medical community, concerns were raised about implementing genetic screening programs even in populations where mutations are more frequent (Healy 1997). The relevance of genetic testing is questioned when the ensuing medical treatment is still unclear. A third factor deals with the psychological and social impacts of screening. On the one hand, there is the issue of the impact of diffusing such information to women and their capacity to correctly interpret it; on the other, there is the issue of privacy protection and insurance industry involvement (Subramanian et al. 1999).

Researcher Strategies

In-depth interviews were conducted to examine how researchers react to the new research environment and how it affects their research practices. The objective was to determine their strategy in dealing with the most important issues facing them: (1) research orientations; (2) organization of research (their research group, the specializations of its members, the roles of students and technicians, collaboration with other groups); (3) financing strategy (resources necessary to be competitive, public financing, private financing); (4) diffusion of information (to the scientific community, to the public); and (5) commercialization of knowledge (industrial applications of their research, commercialization strategies, opinions about patenting).

Researchers were selected from Figure 9.1 so as to cover the different networks and the various interactions with industry. They came from Canada, the United States, and two European countries. Six are affiliated with a public research organization, each having some teaching duty, and one with industry. Each interview lasted between one hour and ninety minutes.[12]

The most significant differences between these researchers, in terms of activities and strategies, are their respective specialties and local and national institutional factors. The main similarity is their strong commitment to the advancement of science as their guiding principle in research. Even the industrial researcher considers this the principal preoccupation, although the organization to which he or she belongs is totally driven by the commercialization of scientific results. All of the researchers interviewed are preoccupied both by the development of clinical applications of their knowledge and its impact on public health and by the role of biotechnology firms in the commercialization of new knowledge. According to the bibliometric analysis, five researchers should have only weak ties with industry, which were corroborated by the interviews, while one academic researcher has frequent interactions with industry. The remaining researcher works in the industry.

Differences in strategies can be explained by individual research backgrounds and specialties. BRCA1 and BRCA2 research in the 1990s attracted researchers with different kinds of expertise and training. Large teams would include experts in breast cancer, immunology, epidemiology, and genetics. Our respondents explained in detail how their contributions, interests, and interactions with other researchers were determined by their specialties. For instance, one researcher, defining himself as an immunologist, stated that his interest centred on the study of hereditary factors associated with breast cancer and on the links between breast cancer genes and the disease. He integrated a larger team researching BRCA1 localization. Owing to his lack of expertise in sequencing techniques, his role at the stage of identification was limited. He therefore felt that his contribution to the development of BRCA1 and BRCA2 research was more significant during certain periods than others, more important in the localization of the genes than in their sequencing. In this context, he naturally found BRCA1 and BRCA2 research to be more interesting during the periods when his specialty coincided with the main interest of his community, and he could then obtain more research funds and publish in *high-impact-factor* journals.[13] This researcher did believe that he would not have obtained adequate research funding and staff had he not collaborated within a large consortium.

Differences can also be explained by local and national institutional factors. The organizations with which the researchers are affiliated control essential research resources. The presence of graduate and postgraduate students and their potential contributions depend on the organization's training programs and rules. In each of their respective research organizations, respondents had a different way of interacting with students. In the case of a researcher directly affiliated with a hospital, the students were mostly physicians specializing in human genetics. In another case, the researcher had no immediate access to students and had to recruit them from other organizations, with profiles that were usually far removed from medical genetics. For five of the seven individuals interviewed, students represented the most important labour force in the laboratory since the number of technicians was very limited in these research centres. One academic researcher seldom hired graduate students for his laboratory because he believed that public financing programs in his country offered only limited support for long-term student training. He preferred to rely on technicians hired on a contractual basis. Otherwise, the industrial researcher in our sample indicated that his firm's policy on the issue was to consider students as laboratory technicians rather than as graduate students. They were given short-term contracts, while at the university they worked on the same research subject throughout their degree programs. By the same token, he believed that industry was not the ideal setting for students because the firm specialized in very risky projects, such as gene sequencing. If the project failed or was stopped, or if another group had priority, the student could encounter difficulty in publishing results in major journals. If it were a success, the firm could delay publication, and the student would not get full credit for its discovery. The university researcher with close ties to industry said that he encountered such problems hiring graduate students on projects financed by the private sector. He subsequently decided to exclude students from such projects. Another respondent also mentioned that technicians and professionals were more suitable than students for such risky projects, where intellectual property and diffusion of information were at stake.

The financing of research activities is also, to a certain extent, decided within the organization, including student working conditions and the financing of equipment. The implication is that the allocation of funds within the organization and in the laboratory is always a strategic issue. Two items are of particular importance. The first is the financing of expensive equipment, which entails a great deal of lobbying to secure needed resources. The second is the allocation of personnel, such as in determining which projects and researchers get the most students. Four researchers complained that their organization was unable to reallocate resources quickly enough over a short period during the race for BRCA1 and BRCA2 sequencing to provide the personnel required to achieve success. For these researchers, the relationship with their respective organizations was very important, and they devoted a large portion of their time to participating in the main decisions.

At the local level, the most important asset was the quality and extent of interaction within a local clinic. Five researchers are affiliated with research centres located in hospitals, and all concluded that this provided a competitive advantage. During the early 1990s, they obtained genetic materials from patients treated in their hospital. A large part of their research was devoted to data analysis. However, one researcher placed less emphasis on his connection with a clinic; when questioned further, he recognized that he nonetheless had access to patients through more official mechanisms. In his laboratory, other members, such as physicians, had direct access to patients and were responsible for providing family data. National organizations responsible for hereditary diseases helped them to obtain access to patients. In all instances, interaction with the clinic and access to genetic materials were a prerequisite to competitiveness. The mechanisms in place depend largely on institutional factors. In each country, testing individuals for genetic disorders depends on the organization of the health care system and on the interactions between patients and the medical community. In each country, rules concerning privacy and confidentiality are different. Two researchers complained that their country enforced more severe rules, that it made it more difficult to establish large databases (see Eischen, this volume, Chapter 6, for an example of how Iceland facilitates the commercialization of this kind of research).

International collaboration and the integration of a large research team increased the competitiveness of research on BRCA1 and BRCA2 genes. The degree to which an organization was integrated into the greater international community was mainly a function of the quality of its genetic materials. Each organization that participated in the International Breast Cancer Linkage Consortium provided its own data. Four respondents mentioned the impact of this consortium, which put together a very large database, including independent data from different national settings. These researchers also said that ad hoc collaboration with other important groups was frequent because they had complementary data or expertise. Such collaboration allowed small national groups, bringing local family-based data, to be associated with large genomic research centres. According to one researcher, an important objective of his organization is to strengthen international collaboration in order to keep pace with the advancement of knowledge and, in projects such as BRCA1 and BRCA2, of new technologies.

The national environment also shaped the clinics themselves and the interaction between researchers and the medical community. Of primary importance were the institutions within the medical establishment and the interaction between actors. The research community as a profession is less regulated than that for physicians and other clinical personnel. For instance, genetics as a specialty was recognized only recently in one country by its association of physicians. The broad issue involves determining which medical specialty will be responsible for genetic testing and counselling. The outcome in turn determines to which medical specialty researchers will direct their attention to relevant issues. Of secondary importance are the different health policies in each country and their impacts on researcher strategies and subsequent commercialization. The diffusion of genetic testing depends both on the reaction of the national and international medical communities regarding positive impacts on health and on health policies implemented by national bodies. The researchers interviewed did not expect the same rate of diffusion in all countries.

Five of the seven respondents said that the private sector played an important role in research activities. The university researcher with strong ties to industry is based in a hospital and believed that he would have been unable to do what he did without the support of industry. He stated, however, that he intended to put more emphasis on public funding in the coming years. According to him, the main advantage of private contracts is related to equipment procurement. Two other researchers also indicated that, whereas research in the early 1990s was done with rather standard facilities, much more sophisticated equipment was required as more researchers entered the race and clinical applications became a key concern.

The weak ties that five of our university researchers have with biotechnology firms can be explained by the national context. In our interviews, strong differences were detectable between European and North American researchers. European researchers are less likely to interact with industry. In both European countries, fewer resources were devoted by local organizations to establishing links with industry, and national policies provide no incentives to foster such interactions and to commercialize new knowledge. One European researcher said that national and local institutional arrangements governing interaction with firms were still in their infancy in his country because the biotechnology industry was weak and not considered a national priority. He believes that foreign biotechnology firms are not willing to establish a subsidiary in his country because they consider that the diffusion of genetic screening is unlikely due to national health policies. Another European researcher, however, mentioned that the trend in national science policies was toward more direct interaction with firms, and he believed that the contribution of university researchers was essential to the development of a strong biotechnology industry.

Three respondents considered that the interaction with firms depends not only on the quality of research and its potential commercial applications but also on the firms' expectations concerning future national markets. In this context, biotechnology firms perceive research contracts with universities as the first step in the commercialization of new health technologies in a country. For most researchers interviewed, such contracts do not significantly contribute to pure research. This statement should be weighted against the views, mentioned earlier, concerning the priority given by all researchers to the advancement of knowledge. Otherwise, most respondents said that interaction with firms will become essential in the future, and, as one mentioned, private-sector contracts can be fruitful if managed properly and governed by formal rules. For example, private firms can contribute to the financing of expensive facilities required to become and remain competitive.

When researchers were asked about which constraints characterized relations with industry, such as lack of autonomy and delays in diffusion of new knowledge, they replied that intellectual freedom was not an issue. Three researchers in the public sector have patented their discoveries, and two precisely in the area of commercial applications of BRCA1 or BRCA2 genes. They explained that their university provided financial and technical support. One acted on a request from his own university, although he himself was rather skeptical about the potential benefits that his organization would gain from his patent. His organization's strategy is to practise counterpatenting against Myriad Genetics in order to increase national bargaining power. Although this American firm has property rights on BRCA1 and BRCA2 applications, four respondents expressed doubts that the genetic screening for such genes would resume and that the firm will reap huge profits from this project.

Finally, researchers were invited to comment on the conflict between Myriad Genetics and two research groups, from the University of Pennsylvania and the British Institute of Cancer Research, about BRCA1 and BRCA2 patents enforcement. There was unexpected unanimity that Myriad Genetics did not have a real advantage compared to its public-sector competitors in scientific research. According to our respondents, the company's financial resources and equipment did not constitute very important assets in the race of BRCA1 and BRCA2 gene sequencing. Collaboration with public organizations was essential to Myriad's success, as was its integration in the international scientific community to gather more information. Competition between teams was, according to three respondents, a driving force and accelerated these scientific discoveries, for which Myriad Genetics filed patents. One respondent even believes that any public organization would have been able to secure priority. As mentioned previously, researchers, however, complained that their own laboratory and university were unable to reorient resources rapidly enough to obtain more skilled labour to pursue BRCA1 and BRCA2 sequencing. Otherwise, four respondents added that the monopoly that Myriad Genetics tried to enforce on both genes seemed to impede neither scientific research on these genes nor its applications. This judgment does not seem to be shared by the two groups officially complaining about Myriad Genetics' patent enforcement. One respondent said that he agrees to pay a "fair" price to Myriad Genetics in royalties, as he does for other technologies. He added that he lobbied elected officials of his country to modify patent laws in order to limit patenting of living organisms.

Conclusion

In the beginning, firms involved themselves in breast and ovarian cancer research as a research organization using in-house scientific resources. Myriad Genetics and its partners were important actors in the race for BRCA1 and BRCA2 gene isolation and identification and were part of the consortium that obtained priority for BRCA1 sequencing and was the runner-up for BRCA2 sequencing. This firm had more traditional scientific interaction with other groups within public organizations through international consortia. Myriad Genetics, as well as other firms, tried to develop commercial applications, the first one being genetic tests, for which Myriad Genetics enforced its patents in the United States and gradually in Europe, but met with a rather limited market. In the area of applications related to treatment, as yet no breakthrough has emerged from this research.

The new research dynamic envisaged by the sociology of science is, however, in its infancy. The diversity of expertise needed in such projects makes transdisciplinary teams an absolute prerequisite. Individual researchers stated clearly that at each stage different disciplines were needed and that their contributions have been more important for certain tasks than for others. Among our respondents and especially in Europe, some teams maintained a traditional scientific organization, hiring almost solely graduate students on the basis of public financing. Two respondents have adopted different research dynamics. The researcher from industry does not participate in student training yet is in indirect and frequent contact with university researchers, so that in practice frontiers between organizations are blurring, despite an apparent lack of involvement of graduate students in the firm. The university researcher collaborating with industry said that projects financed by industry and those by research councils are separated and carried out in a different research dynamic. Projects involving private funds generally excluded graduate students. Other public-sector respondents believe that industry will be a more important actor in the future. Its role depends on institutional arrangements, such as its weight in national science policies and the structures governing university-industry relations within the organization.

National differences are still important. The literature on biotechnology recognizes this when discussing the impacts of national institutional arrangements. However, scholars associated with the sociology of science concentrate attention more specifically on the North American context, for which their findings are probably more attuned. According to our interviews, European research organizations adopt a different research dynamic than their North American counterparts, at least for commercialization of research and for institutional arrangements regulating university-industry interactions.

As expected, intellectual property was a major issue in BRCA1 and BRCA2 research, and several organizations filed patents to protect and commercialize their discoveries. Myriad Genetics secured rights on a large number of patents in order to position itself as the apparent leader in industrial and clinical applications. This company is currently trying to enforce this in spite of opposition from the scientific community. Our respondents demonstrated a rather moderate opinion on this issue, probably because they believe that in the short term the market is and probably will remain rather limited. Our data suggest that scientific collaboration is more difficult when intellectual property is at stake. For instance, Myriad Genetics' research network seems to be less interorganizational than its main competitor, the European network.

Studies on biotechnology insist on the need for scientific and industrial networks for successful commercialization. According to our interviews, some actors in large networks are more important than anticipated by this literature; for instance, the broad medical network was a key player in the diffusion of applications to patients. The rather cool reaction by the BRCA1 and BRCA2 community and the medical community on the subject of genetic screening delayed testing. A lack of consensus about the role of genes in these diseases and about the accuracy and interpretation of genetic tests hindered the sale of diagnostic tests. In scientific research per se, interaction with a clinic was also important. Most researchers mentioned that access to patients was crucial to building reliable family databases; the strategy that they chose to reach this goal depended on their links to a clinic and the national regulation of relations between researchers and patients. In turn, their integration in international scientific networks depended partly on such data.

As in any case study, our results should not be expanded to cover the whole spectrum of biotechnology research. The advantage of case studies is that they offer an in-depth analysis of one area, so that their specificity becomes more obvious. For instance, conflicts over intellectual property were probably intensified by high market expectations when genes were discovered. Furthermore, since it was the first gene identified for a frequent and severe disease, most actors also took into consideration its potential impact on other health technologies. One of the most important issues not raised in this chapter is the social benefits of public investments in biotechnology. Measuring this factor appears to be very difficult, since benefits should be evaluated from a public health point of view. Only in the long run will it be possible to show whether such research has contributed to the treatment of breast and ovarian cancer.

Acknowledgments

A preliminary version of this chapter was published in Robert Dalpé, Louise Bouchard, and Daniel Ducharme, "Scientific, Medical, and Industrial Issues in Breast and Ovarian Cancer Genes Research," *University as a Bridge from Technology to Society: Proceedings of the IEEE International Symposium on Technology and Society*, 91-99. New Brunswick, NJ: IEEE, 2000. This research project was funded by SSHRCC.

Notes

1 Monoclonal antibodies are highly specific and can be used to diagnose disease.

2 As we will see later, successful commercial application will encounter various obstacles, requiring important adjustments in institutional arrangements governing research and the commercialization of biotechnology. For instance, changes were made to rules pertaining to the patenting of living organisms.

3 According to a 2001 list of top fifty biological firms compiled by *Genetic Engineering News,* Genentech is number three, with a market capitalization of $26 billion US.
4 According to Polanyi (1962), the social system of the Republic of Science implies that researchers choose research subjects based on intellectual curiosity. Individual benefits will flow from prestige and the social importance of their function. Scientific disciplines are self-regulated, meaning that less efficient researchers or those working on subjects perceived as nonfruitful will be excluded.
5 According to this model, links are more intense between these three actors. In this respect, universities will obtain an additional function, contributing to industrial innovation and economic growth. They should sustain entrepreneurship, especially in interaction with other local actors.
6 This section is partly derived from Dalpé, Bouchard, Houle, and Bédard (2003).
7 These publications were indexed by the Sciences Citation Index and the Social Sciences Citation Index. The search was done in the title of publications in these two databases for the keywords *BRCA, BRCA1, BRCA2,* or *BRCA3.* Standardization of the label BRCA occurred in 1993. These 803 publications represent a rather tightly knit environment since almost all papers most highly cited by the 803 publications are also in the database.
8 In order to prevent the diffusion of information by peer reviewers to competing teams, the paper submitted to *Science* did not include the complete sequence, which the authors added only when the paper was published.
9 Factor analysis is here obtained by the matrix of coauthorship of the 803 publications for the fifty researchers having the largest number of publications in the bibliometric database.
10 They wrote papers with researchers from almost all networks.
11 The cost of Myriad's test is $2,500 US.
12 The rules of confidentiality established in our research institutions are enforced.
13 His words.

References

Abraham, C. 2003. "Ontario to Fight for Gene Test." *Globe and Mail,* 7 January, A5.

Argyres, N.S., and J.P. Liebeskind. 1998. "Privatizing the Intellectual Commons: Universities and the Commercialization of Biotechnology." *Journal of Economic Behaviour and Organization* 35: 427-54.

Armstrong, K., A. Eisen, and B. Weber. 2000. "Primary Care: Assessing the Risk of Breast Cancer." *New England Journal of Medicine* 342 (8): 564-71.

Arver, B., Q. Du, J. Chen, L. Luo, and A. Lindblom. 2000. "Hereditary Breast Cancer: A Review." *Seminars in Cancer Biology* 10: 271-88.

Audretsch, D.B., and P.E. Stephan. 1996. "Company-Scientist Locational Links: The Case of Biotechnology." *American Economic Review* 30: 641-52.

Benzie, R. 2001. "Ontario to Defy U.S. Patents on Cancer Genes." *National Post,* 20 September, A15.

Blumenthal, D., E.G. Campbell, M.S. Anderson, N. Causino, and K.S. Louis. 1997. "Withholding Research Results in Academic Life Science: Evidence from a National Survey of Faculty." *JAMA* 277: 1224-28.

Bush, V. 1980 [1945]. *Science: The Endless Frontier.* New York: Arno Press.

Butler, D., and D. Gershon. 1994. "Breast Cancer Discovery Sparks New Debate on Patenting Human Genes." *Nature* 371: 271-72.

Butler, D., and S. Goodman. 2001. "French Researchers Take a Stand against Cancer Gene Patent." *Nature* 413: 95-98.

Buttler, D. 2000. "US/UK Statement on Genome Data Prompts Debate on 'Free Access.'" *Nature* 404: 324-25.

Cassier, M., and D. Foray. 1999. "La Régulation de la propriété intellectuelle dans les consortiums de recherche: Les Types de solutions élaborées par les chercheurs." *Économie appliquée* 52: 155-82.

Cho, M.K., R. Shohara, A. Schissel, and D. Rennie. 2000. "Policies on Faculty Conflicts of Interest in US Universities." *JAMA* 284: 2203-8.

Dalpé, R., L. Bouchard, A.J. Houle, and L. Bédard. 2003. "Watching the Race to Find the Breast Cancer Genes." *Science, Technology, and Human Values* 28 (3): 187-216.

Dalpé, R., and M.P. Ippersiel. 2000. "Réseautage et relations avec l'industrie dans les nouveaux matériaux et l'optique." *Sociologie et sociétés* 32 (1): 107-34.

Davies, K., and M. White. 1996. *Breakthrough: The Race to Find the Breast Cancer Gene.* New York: John Wiley and Sons.

Deeds, D.L., D. Decarolis, and J.E. Coombs. 1997. "The Impact of Firm-Specific Capabilities on the Amount of Capital Raised in an Initial Public Offering: Evidence from the Biotechnology Industry." *Journal of Business Venturing* 12: 31-46.

Easton, D.F., D.T. Bishop, D. Ford, and G.P. Crockford. 1993. "Genetic Linkage Analysis in Familial Breast and Ovarian-Cancer: Results from 214 Families." *American Journal of Human Genetics* 52 (4): 678-701.

Edquist, C., ed. 1997. *Technologies, Institutions, and Organizations.* London: Frances Pinter.

Elwood, J.M. 1999. "Santé publique et dépistage génétique du cancer du sein au Canada – première partie: Risques et interventions." *Maladies chroniques au Canada* 20 (1): 4-16.

Etzkowitz, H., and L. Leydesdorff. 1997. "The Dynamic of Innovation: From National Systems and 'Mode 2' to a Triple Helix of University-Industry-Government Relations." *Research Policy* 29: 109-23.

Gad, S., et al. 2001. "Identification of a Large Rearrangement of the BRCA1 Gene Using Colour Bar Code on Combed DNA in an American Breast/Ovarian Cancer Family Previously Studied by Direct Sequencing." *Journal of Medical Genetics* 38: 388-92.

Gauthier, É. 1998. *L'Analyse bibliométrique de la recherche scientifique et technologique: Guide méthodologique d'utilisation et d'interprétation.* Ottawa: Statistique Canada.

Gibbons, M., C. Limoges, H. Nowotny, S. Schwartzman, P. Scott, and M. Trow. 1994. *The New Production of Knowledge: The Dynamics of Science and Research in Contemporary Societies.* London: Sage Publications.

Giesecke, S. 2000. "The Contrasting Roles of Government in the Development of Biotechnology Industry in the US and Germany." *Research Policy* 29: 205-23.

Godin, B. 1998. "Writing Performative History: The New New Atlantis?" *Social Studies of Science* 28: 465-83.

Granovetter, M. 1985. "Economic Action and Social Structure: The Problem of Embeddedness." *American Journal of Sociology* 91 (3): 481-510.

Hall, J.M., M.K. Lee, B. Newman, J.E. Morrow, L.A. Anderson, B. Huey, and M.C. King. 1990. "Linkage of Early-Onset Familial Breast-Cancer to Chromosome 17Q21." *Science* 250: 1684-89.

Healy, B. 1997. "BRCA Genes – Bookmaking, Fortunetelling, and Medical Care." *New England Journal of Medicine* 336: 1448-49.

Heilbron, J.L., and W.F. Bynum. 2000. "Millennial Highlights ... from Gerbert d'Aurillac to Watson and Crick." *Nature* 403: 13-16.

Hurst, L. 2001. "U.S. Firm Calls Halt to Cancer Test in Canada." *Toronto Star,* 11 August, A1.

Kenney, M. 1986. *Biotechnology: The University-Industrial Complex.* New Haven: Yale University Press.

Kogut, B. 1989. "The Stability of Joint Ventures: Reciprocity and Competitive Rivalry." *Journal of Industrial Economics* 38: 183-98.

Kotulak, R. 1999. "Biotech Firms Say They Need Protection." *Chicago Tribune,* 12 September, 1.

Lewis, R. 2000. "U.S.-U.K. Joint Statement Creates Initial Stir among Commercial Interests." *Scientist* 14: 1.

Meek, J. 2000. "Money and the Meaning of Life." *Guardian,* 17 January, 6.

Miki, Y., et al. 1994. "A Strong Candidate for the Breast and Ovarian-Cancer Susceptibility Gene." *Science* 266: 66-71.

Mowery, D.C., R.R. Nelson, B.N. Sampat, and A.A. Ziedonis. 2001. "The Growth of Patenting and Licensing by U.S. Universities: An Assessment of the Effects of the Bayh-Dole Act of 1980." *Research Policy* 30 (1): 99-119.

Oberbeck, S. 1996. "Deny S.L. Firm a Patent, Groups Urge Breast-Cancer Studies Will Suffer, They Say Deny Firm's Patent, Groups Urge." *SL Tribune,* 22 May, B4.

Pisano, G.P., W. Shan, and D.J. Teece. 1988. "Joint Ventures and Collaboration in the Biotechnology Industry." In D.C. Mowery, ed., *International Collaborative Ventures in U.S. Manufacturing,* 183-222. Cambridge, MA: Ballinger.

Polanyi, M. 1962. "The Republic of Science: Its Political and Economic Theory." *Minerva* 1: 54-73.

Porter, M.E. 1998. "Clusters and the New Economics of Competition." *Harvard Business Review* November-December: 77-90.

Powell, W.W., and J. Owen-Smith. 1998. "Universities and the Market for Intellectual Property in the Life Sciences." *Journal of Policy Analysis and Management* 17 (2): 253-77.

Powell, W.W., K.W. Koput, L. Smith-Doerr, and J. Owen-Smith. 1999. "Network Position and Firm Performance: Organizational Returns to Collaboration in the Biotechnology Industry." *Research in the Sociology of Organizations* 16: 129-59.

Prevezer, M. 1997. "The Dynamics of Industrial Clustering in Biotechnology." *Small Business Economics* 9: 255-71.

Rosenberg, N., and R.R. Nelson. 1994. "American Universities and Technical Advance in Industry." *Research Policy* 23 (3): 323-48.

Ruivo, B. 1994. "'Phases' or 'Paradigms' of Science Policy?" *Science and Public Policy* 21 (3): 157-64.

Scully, R., and D.M. Livingston. 2000. "In Search of the Tumor-Suppressor Functions of BRCA1 and BRCA2." *Nature* 408: 429-32.

Shohet, S., and M. Prevezer. 1996. "UK Biotechnology: Institutional Linkages, Technology Transfer, and the Role of Intermediaries." *R&D Management* 26 (3): 283-98.

Slaughter, S., and L.L. Leslie. 1997. *Academic Capitalism: Politics, Policies, and the Entrepreneurial University.* Baltimore: Johns Hopkins University Press.

Struewing, J.P., P. Hartge, S. Wacholder, S.M. Baker, M. Berlin, M. McAdams, M.M. Timmerman, L.C. Brody, and M.A. Tucker. 1997. "The Risk of Cancer Associated with Specific Mutations of BRCA1 and BRCA2 among Ashkenazi Jews." *New England Journal of Medicine* 336: 1401-8.

Subramanian, K., J. Lemaire, J.C. Hershey, M.V. Pauly, K. Armstrong, and D.A. Asch. 1999. "Estimating Adverse Selection Costs from Genetic Testing for Breast and Ovarian Cancer: The Case of Life Insurance." *Journal of Risk and Insurance* 66: 531-50.

Tavtigian, S.V., et al. 1996. "The Complete BRCA2 Gene and Mutations in Chromosome 13Q-Linked Kindreds." *Nature Genetics* 12 (3): 333-37.

UK Clinical Molecular Genetics Society. 1999. *Gene Patents and Clinical Molecular Genetic Testing in the UK: Threats, Weaknesses, Opportunities, and Strengths.* Cambridge, UK: Public Health Genetics Unit.

Van Raan, A.F.J., ed. 1988. *Handbook of Quantitative Studies of Science and Technology.* Amsterdam: North Holland.

Wooster, R., et al. 1994. "Localization of a Breast-Cancer Susceptibility Gene BRCA2 to Chromosome 13Q12-13." *Science* 265: 2088-90.

–. 1995. "Identification of the Breast-Cancer Susceptibility Gene BRCA2." *Nature* 378: 789-92.

Zucker, L.G., M.R. Darby, and M.B. Brewer. 1998. "Intellectual Human Capital and the Birth of U.S. Biotechnology Enterprises." *American Economic Review* 88 (1): 290-306.

Contributors

Louise Bouchard received her PhD from the University of Montreal in the sociology of health and a postdoctoral scholarship from INSERM-U379 in Marseilles, France. She teaches in the Department of Sociology and in the PhD program of population health at the University of Ottawa. Her research interests include the social impact of technological change involved in the development of prenatal diagnosis, in predictive testing for cancers, and in the controversial aspects of the diffusion of molecular tests in clinical practice. In parallel, she has a deep interest in population health issues, a topic that she teaches and in which she has begun to develop research projects (e.g., being in a minority group and health status). She is also involved in the study of social health determinants, with a particular interest in social capital.

Jacqueline E.W. Broerse graduated in medical biology at the Vrije Universiteit Amsterdam (1988) and wrote her PhD dissertation on the development of a participatory approach to include small-scale farmers in the biotechnological innovation process at the Vrije Universiteit Amsterdam (VUA). Since 2000, she has been an associate professor at the Institute for Innovation and Transdisciplinary Research of the VUA. After her graduation, she worked for several years as a policy maker on biotechnology at the Department of Research and Technology of the Dutch Directorate General for International Cooperation (Ministry of Foreign Affairs). She then became a lecturer in the Department of Biology and Society as well as a project officer at the Centre for Development Cooperation Services, both of the VUA. She is a member of the Platform Integrated Plant Conversion of the Dutch Foundation on Sustainable Chemistry Development (Stichting Duurzame Chemie Ontwikkeling). Her current research is focused on methodology development for interactive policy and interactive technology development in general and transdisciplinary knowledge integration in particular.

Joske F.G. Bunders graduated in chemistry and physics at the Universiteit van Amsterdam and wrote her PhD dissertation on participatory approaches to the development of science-based innovations in agriculture at the Vrije Universiteit Amsterdam (1994). Since 2002, she has been the director of the Institute for Innovation and Transdisciplinary Research of the VUA. She was appointed professor of biology and society at the Vrije Universiteit Amsterdam in 2000. She participates in several governmental committees and advisory boards, such as the Gender Advisory Board of the United Nations Commission on Science and Technology for Development (UNCSTD), the Committee on Technology Assessment of the Dutch Ministry of Agriculture, Nature, and Fisheries, and the Dutch advisory council for research on spatial planning, nature, and the environment (RMNO). She is vice chair of the Committee on Nature and Health (subcommittee of the Dutch Health Council). Her specific field of interest is the linking of knowledge and expertise of end users (e.g., small-scale farmers or patients) with developments in modern science and (inter)national policy.

Annette Burfoot is an associate professor in the Department of Sociology, Queen's University, where she teaches and researches science and technology studies. She adopts a feminist and sometimes historical perspective in the cultural study of science and technology. Her major publications include *The Encyclopedia of Reproductive Technologies* (Westview, 1999) and *Killing Women: The Visual Culture of Gender and Violence* (Wilfrid Laurier University Press, forthcoming). She has published many articles on new reproductive technologies, genetic engineering, and biotechnology. Recently, she has been working on early modern medical visualization technologies (eighteenth-century wax anatomical models) as significant precedents for contemporary medical visual culture. She is also working on a collaborative project with Jennifer Poudrier on medical science studies.

Robert Dalpé is an associate professor of political science at L'Université de Montréal and a researcher at the Center for Interuniversity Research on Science and Technology (CIRST). His research projects deal with science policies and university-industry relations in genetics and biotechnology.

Daniel Ducharme received his PhD in sociology from L'Université de Montréal. He was a research assistant at the Centre de recherche en droit public, Institut de recherches cliniques de Montréal, and Groupe ÉCOBES. He is a postdoctoral fellow at the Center for Interuniversity Research in Science and Technology (CIRST) in Montreal, studying the medical, social, and ethical dynamic of genetic testing in Quebec.

Kyle Eischen is the associate director of Regional and Informational Research at the Center for Global, International, and Regional Studies, University of California at Santa Cruz. His research centres on information, technology, and social impacts, with a focus on work, organization, and regional economic development. Currently, he is a lead investigator of the evolution of Silicon Valley from "bits to genes" and the corresponding global links between the United States, India, and China in software and biotechnology. Previously, he was a Reuters Digital Vision Fellow at Stanford University (2002-3) and the cofounder of the Abrivo Foundation. He holds a doctorate in sociology from the University of California, Santa Cruz, and a master's in applied economics and international affairs from the Graduate School of International Relations and Pacific Studies at the University of California at San Diego.

Neil Gerlach is an assistant professor of sociology in the Department of Sociology and Anthropology, Carleton University. He holds degrees in sociology, anthropology, law, and education and teaches in the areas of social theory, biotechnology in criminal justice, and organization studies. His research interests revolve around the question of the roles of technology in changing forms of social governance, and he has published in the areas of biotechnology policy, management discourse, and gender and technology. He is the author of *The Genetic Imaginary: DNA in the Canadian Criminal Justice System* (University of Toronto Press, 2004).

Craig K. Harris is a faculty member of the Department of Sociology at Michigan State University, where he is also appointed in the Michigan Agricultural Experiment Station, the National Food Safety and Toxicology Center, and the Institute for Food and Agricultural Standards. In addition to conducting research on biotechnology policy and perception, he has studied perception and policy with respect to pest management in fruit and vegetable production and the management of changing fisheries in the Great Lakes of North America and sub-Saharan Africa. His emerging interests include the role of biotechnology in agave cactus production and the development of standards for organic fisheries.

Michael D. Mehta specializes in science, technology, and society, with a particular interest in health and environmental policy. His interests encompass a range of risk domains, including biotechnology, nanotechnology, microsystems, nuclear safety, climate change, blood safety, et cetera. His academic background includes a BA in psychology, a master's in environmental studies, a PhD in sociology, and postdoctoral training in policy studies. He has held academic appointments at York University (Faculty of Environmental Studies) and Queen's University (School of Policy Studies and School of Environmental Studies), and he has taught graduate and undergraduate students for fifteen years. He is a cofounder of the Environmental Studies Association of Canada (ESAC). Dr. Mehta is an associate professor of sociology at the University of Saskatchewan and chair of the Sociology of Biotechnology Program. He is also the director of the Social Research Unit. Finally, Dr. Mehta sits on the board of directors for the provincial electrical utility SaskPower.

Jennifer Poudrier earned her PhD from Queen's University in Kingston. Currently, she is an assistant professor in the Department of Sociology at the University of Saskatchewan and teaches in the areas of medical sociology, medical science studies, women's health, and the sociology of aging. Her current research interests lie at the intersections between medical sociology, Aboriginal knowledges, and science studies in the areas of medicine, genetics, and nanoscience. Currently, she is involved in several research projects related to Aboriginal health in Canada and new genetic science. She is also working on a collaborative project with Annette Burfoot concerned with the development of medical science studies.

Toby A. Ten Eyck is an assistant professor in the Sociology Department and National Food Safety and Toxicology Center at Michigan State University. His work focuses on media presentations and audience interpretations of food safety issues, such as irradiation, biotechnology, and food-borne pathogens. His work can be found in such journals as *Rural Sociology, Science Communication, Society,* and *Journal of Commercial Biotechnology.*

Christopher K. Vanderpool (1943-2001) earned his PhD from Michigan State University in 1970 and returned to the faculty in 1974, where he became chair of the Sociology Department in 1990, holding that position until 1999. His work took him to Africa, Asia, and Latin America, focusing mainly on rural issues such as land use and farming techniques. As his obituary in the American Sociological Association's *Footnotes* said, "His collaborators regarded him as an ideal partner in scholarly activity, always curious, intellectually alert, and scrupulously objective."

Margareta Wandel holds a PhD in public nutrition and is a professor in the Department of Nutrition at the Institute for Basic Medical Sciences, University of Oslo, Norway. Her research activities revolve around changes in food habits and the implications for health in different groups of the population. Her focus is on the mechanisms for this change, particularly on the role of the socio-economic and cultural environments in shaping patterns of health-related behaviour. A special interest is how information about food is perceived and acted on by consumers.

Index

Note: "ag-biotech" stands for agricultural biotechnology; GM, for genetically modified; GMO, for genetically modified organism; IPR, for intellectual property rights

Printed and bound in Canada by Friesens

Set in Stone by Artegraphica Design Co. Ltd.

Copy editor: Dallas Harrison

Proofreader: Deborah Kerr

Indexer: Patricia Buchanan